Digital Economies:
SMEs and E-Readiness

Stephen M. Mutula
University of Botswana, Botswana

BUSINESS SCIENCE REFERENCE
Hershey · New York

Director of Editorial Content: Kristin Klinger
Senior Managing Editor: Jamie Snavely
Assistant Managing Editor: Michael Brehm
Publishing Assistant: Sean Woznicki
Cover Design: Lisa Tosheff
Printed at: Yurchak Printing Inc.

Published in the United States of America by
Business Science Reference (an imprint of IGI Global)
701 E. Chocolate Avenue
Hershey PA 17033
Tel: 717-533-8845
Fax: 717-533-8661
E-mail: cust@igi-global.com
Web site: http://www.igi-global.com/reference

Library of Congress Cataloging-in-Publication Data

Mutula, Stephen M.

Digital economies : SMEs and e-readiness / by Stephen M. Mutula.
p. cm.
Includes bibliographical references and index.
Summary: "This book provides valuable insights into the current state of the digital economy and the ability of SMEs to leverage Information and Communication Technologies (ICTs) so that they may overcome their traditional laggard position in the global business market"--Provided by publisher.
ISBN 978-1-60566-420-0 (hardcover) -- ISBN 978-1-60566-421-7 (ebook) 1. Small business. 2. Electronic commerce. 3. Technological innovations. I. Title.

HD2341.M88 2009
381'.142--dc22

British Cataloguing in Publication Data
A Cataloguing in Publication record for this book is available from the British Library.

Table of Contents

Section 5:
Content, Knowledge, and Records Management

Section 6:
Information Needs and Capacity Building in SMEs

Section 7:
Globalisation, Trends, and Best Practices in the Digital Economy

Foreword

It is a new world. In the old world, industrialization was a *sine qua non* to development and poverty alleviation or eradication. Countries needed to possess or acquire natural resources, transform them through processing into higher value manufactured products, and export them to other countries. This was the way the first world attained its preeminence in industrialisation and growth. That was the model of development that made many newly independent countries in Africa and Asia adopt industrialization as the route to economic development and growth. Apart from the low level of technological knowledge and skills that constrained the success of this model, there were problems of infrastructural inadequacy, poor road networks, inadequate electricity, foreign market inaccessibility, and low levels of marketing knowledge.

In the new world, the paradigm has shifted. The most prominent factor that has contributed to this shift has and continues to be advances in technology, particularly in information and communications technologies (ICTs), which have made instant communication possible, thereby facilitating and accelerating the globalisation of the world. The advances in ICT created a new industry, opened new vistas of opportunities in knowledge, and changed the approach in many areas of human endeavour (e.g., business, healthcare, manufacturing, government, education, architecture, services, etc.). In the process, ICT has also expanded the concept of infrastructure beyond roads, ports, and rail lines to include e-infrastructure (i.e., access to computer facilities, availability of broadband Internet connectivity, and the development of human knowledge). It is doubtful whether any country can aspire or attain development and poverty alleviation in the new world without embracing ICT.

Two young Americans, Sergey Brin and Larry Page, created Google, "the biggest media company in the world" (Time, 2006), not by processing any raw material into any physical product, but by creating a product of the mind, a service to store, find, and retrieve information. While still studying for his degree, another young American, 20-year old Shawn Fanning wrote file-swapping music software which was commercialised into a music exchange service that gained over 80 million subscribers in two years (O'Connor and Solomons, 2001). This is the new face of global exports. They depend less on natural or physical capital, or even experience, and more on knowledge and ideas.

Jeffrey Sachs (2005) attributed the cause of extreme poverty in India to a lack of six major kinds of capital, one of which is "knowledge capital: the scientific and technological know-how that raises productivity in business output and the promotion of physical (roads, power, water and sanitation, airports and seaports, and telecommunications systems) and natural (arable land, healthy soils, biodiversity, and well-functioning ecosystems) capital." The focus on the development of "knowledge capital" in India has demonstrated an alternative route to economic development and poverty alleviation in developing countries. While India's textile industries still record substantial exports, it is the export of the new information technologies that has driven the recent rapid growth of the country's economy. India's massive investment in technological education has produced first class engineers and information technology scientists who dominate the ICT service industry and who have availed themselves to the opportunities provided by the liberalisation of the Indian economy to set up ICT service industries in India, to which several large western corporations outsource their services, particularly labour intensive services. The remittances of Indian workers abroad and the increased employment opportunities at home have substantially helped to promote growth and reduce poverty.

What is of interest here is the nexus between the new technologies and economic development. In many economies (both developed and developing), small and medium-sized enterprises (SMEs) account for a disproportionately large contribution to economic growth and development. In general, SMEs are more labour-intensive, depend more on local materials, and utilise low level technologies that are easily accessible to a larger population. They therefore provide greater opportunities for more people to earn an income and leave the poverty zone. However, SMEs, particularly those in developing countries, often suffer from a number of constraints. Their operations are often limited to domestic markets, and they often do not have adequate resources or skills to explore and exploit international markets. What Professor Mutula has done in this book, "*Digital Economies: SMEs and E-Readiness,*" is to analyse the opportunities provided by ICT knowledge to develop SMEs and enable them to function in an increasingly globalised, 'technomanic' world. The book adopts a development and global focus. It summarises the developments and growth of the digital economy and its components, and analyses the readiness of the various sectors and geographical regions of the world before focusing on the applications of e-knowledge to government, commerce, industries, records management, and so forth.

The detailed consideration of the growth potential and challenges of SMEs in availing themselves to the opportunities provided by ICT is of special interest.

The policy recommendations on capacity building in SMEs and the examples of best practices in the developed world present a road map for SMEs in developing countries to transform themselves into greater participants in the global economy, thereby leveraging the pace of development and accelerating the rate of poverty reduction. In "*Digital Economies: SMEs and E-Readiness,*" Professor Mutula has presented ideas, challenges, and policy changes, which, if accorded the seriousness they deserve, can "cheetah pole-vault" (Spio-Garbrah, 2008) the e-readiness of the third world and its crucial SME sector.

Olukunle Iyanda
Professor in Marketing and International Business
University of Botswana
Gaborone (2008)

REFERENCES

O'Connor, A., & Solomons, M. (2001). Spinning closer to the edge. *Financial Times, 3*. In G. Hooley, et al. (Eds.). (2004), *Marketing strategy and competitive positioning*. London: Prentice Hall.

Sachs, J. (2005). *The end of poverty*. London: Penguin.

Spio-Garbrah, E. (2008). *We must fully tap into the information and knowledge age*. Retrieved on November 29, 2008, from *http://www.ngrguardiannews.com/Sunday_magazine/article01//indexn2_html?pdate=231*

Time. (2006). *In search of the real Google*. Retrieved on November 28, 2008, from *http://www.time.com/time/magazine/article/0,9171,1158961,00.html*

Olukunle Iyanda *received his undergraduate education in the University of Lagos and his graduate studies from NYU Stern School of Business. Currently a professor in the Department of Marketing of the University of Botswana, Prof Iyanda previously taught at the University of Lagos, Nigeria, where he also served as Dean of the Faculty. He served as the Founding Dean of Business at Africa University, Zimbabwe and as Director-General of the Nigerian Institute of Management. Prof Iyanda has edited four books and his articles have appeared in such journals as New Academy Review, Management Decision, African Journal of Business, Nigerian Journal of Economics and Social Sciences, and the International Journal of Emerging Markets*

Preface

Several factors motivated the writing of "*Digital Economies: SMEs and E-Readiness*." For one, the dual importance of e-readiness and small and medium-sized enterprises (SMEs) in the digital economy cannot be overemphasized. SMEs are vital for the growth and innovation of dynamic economies, particularly because they diversify national economies while generating employment. For SMEs to play their rightful role in the emerging digital economy–driven by the evolution in ICTs, the Internet, and the World Wide Web–-they need to have attained some reasonable level of e-readiness (i.e., their preparedness to partake in the global information economy based on their capacity to access and use information, access requisite technological infrastructure, have in place adequate human resources, and operate in an enabling legal and regulated business environment).

This book aims to provide valuable insights into the current state of the digital economy and the ability of SMEs to leverage information and communication technologies (ICTs) so that they may overcome their traditional laggard position in the global business market. The topics discussed should carry favour with most stakeholders in the SME industry, such as managers of SMEs, policy makers in government and public administration (the main change agents in the adoption and diffusion of Internet commerce among SMEs), researchers who wish to develop interventions, models, or theories to help explain SMEs in the digital economy, and students in various disciplines-business, information systems, information technology, information science, info-preneurship, and information management-who need material that comprehensively covers the three core areas discussed in this book for learning purposes.

Although several articles and technical reports have been written on e-readiness and SMEs in general, little treatment has been extended to cover the e-readiness of SMEs with regard to their ability to participate in the digital economy. The strengths of this book are in its holistic and inextricable approach to the treatment of the three core subjects of SMEs, digital economies, and e-readiness, and the theoretical and practical flavour of the treatment of the subjects and presentation of cases, experiences, and best practices. Examples that illustrate scenarios are drawn cross-jurisdictionally from both developed and developing countries. This approach is considered vital given the rapid globalisation and digitisation of the business environment, and the pressing need for more information and knowledge to determine the patterns of cross-national diffusion of technology in different cultural settings. Thus, the separate but inextricable relationships of SMEs, digital economies, and e-readiness, which most books treat in isolation, are coalesced into this single book.

The content of this book comes at a time when there is a shortage of information and sources dealing with SMEs, digital economies, and e-readiness. The global e-government and e-readiness reports for 2008 (as supplied by the United Nations and the Economist Intelligence Unit/IBM Institute for Business Value, respectively) do not cite any books in their list of references. Most of the references listed are technical reports, commissioned studies' Web-based sources, and journal articles. The emerging digital economy is just beginning to be appreciated. Consequently, the significant positive effects that the Internet and e-commerce will have on the traditional value chain and business processes call for a reassessment of the traditional roles of SMEs in society. This is in order to design interventions that would enable them to enhance their e-readiness so that they may benefit from the globalized digital economy. Although the Internet is believed to be an important tool for commercial and consumer transactions, only a small proportion of SMEs in most economies have adopted the technology. It is important for all the various factors that militate against SMEs to be interrogated and unpacked in order to help policy makers and SMEs' owners find solutions that would help position these enterprises in leading roles in the digital economy. This book provides a framework for scholars and governments to reassess models that would enhance the capacity of SMEs in the digital dispensation. The book is presented in an easy to understand scholarly fashion, using tables to summarize complex data.

Despite the strengths of the book as stated above, it is possible that the examples and best practices used to illustrate scenarios are biased towards developed countries, which have made great strides in the subjects discussed. The topics covered here are of great international interest. Quite possibly, some of the issues covered in this book may seem out of date by the time this book is published; nevertheless, the book will still reflect the global growth, dynamics, and trajectories of the subjects discussed over time. Tracking the development of a subject or discourse can be of significant importance to educators and students. Furthermore, because the subjects of SMEs, digital economies, and e-readiness are still nascently evolving, most sources one comes across on these subjects are electronic and often Web-based. Although the book has relied on Web resources, most of the sources used are from formal e-journals, repositories, and informal scholarly communication. Increasingly, a number of Web publications now maintain high standards of quality since they are subject to peer review processes. Web-based resources offer researchers and institutions easy access to scientific knowledge, particularly to those located in developing countries, and also for third parties such as economic, business, industrial, and political groups. The sources that are used in this book from the Web have been carefully selected to ensure that they are based on authoritative studies of reputable organisations and/or individuals.

In spite of the great efforts that have gone into the writing of this book to ensure that it comprehensively covers the subjects discussed, it is imprudent to assume that any book can cover all aspects of such topical and dynamic subjects as SMEs, digital economies, and e-readiness. The scope of the book is therefore limited to the interplay and relationships between SMEs, digital economies, and e-readiness. The technical aspects of digital economies, such as economic models and theories underpinning digital economies, are not covered. The focus here lies on operational, management, and policy issues, digital economic components, digital economy applications and processes (e-commerce, e-business, knowledge management, e-records management, etc), e-readiness assessments (of SMEs and macro enterprises), challenges of SMEs in the digital economy and how they can be ameliorated, capacity building of SMEs, and best practices in digital economies.

This book has significant policy and managerial implications. It provides vital information to governments and development agencies on broad-based issues relating to SMEs, digital economies, and e-readiness that can assist them in developing strategic objectives and legal and regulatory frameworks that promote economic growth, infrastructure development, capacity building, and education and training, create awareness of SMEs and the digital economy, and enable access for SMEs to best practices and benchmarking, among other factors. Moreover, SMEs, digital economies, and e-readiness have many dimensions that are of a structural (organisational factors), environmental (digital environment), and technological (e-readiness) nature. Consequently, the managerial understanding of these issues is necessary to assist SMEs to play a leading role in the digital economy. With regard to e-readiness assessments, these are useful information gathering mechanisms that would assist governments when planning strategies for ICT integration, understanding and identifying key and relevant ICT-based development opportunities, and developing enabling strategies and action plans that would address the opportunities and constraints of leveraging ICTs for business competitiveness.

Acknowledgment

The completion of this book, Digital Economies, SMEs and E-readiness, wouldn't have been possible without a considerable assortment of input and institutional and individual support. My thanks are due to the University of Botswana's Vice-Chancellor, Prof. Bojosi Otlhogile, who granted me leave for ten months in 2008 to proceed with my sabbatical at the Department of Library and Information Science, University of Zululand (South Africa). Similarly, my thanks are conveyed to Prof. Frank Youngman - Deputy Vice Chancellor, Academic Affairs, University of Botswana - who has always been an inspiration in my professional life. His parting words on the commencement of my sabbatical leave were, "I have a lot of expectations from you!"

This book was largely completed during my sabbatical sojourn at the Department of Library and Information Science at the University of Zululand. I wish therefore to extend my sincere thanks to Prof. Dennis Ocholla for inviting me to serve as a visiting professor at the university. Prof. Ocholla provided me with tremendous support, especially by facilitating my access to the Internet both at work and at home, and also by involving me in various professional engagements within the country, which subsequently enriched the content of this book. He regularly read through my chapters and offered advice, and ensured that I had access to transport when I needed it throughout my stay. His support at both professional and individual level was therefore inspiring. It would also be remiss of me at this point if I did not recognize the hospitality I enjoyed from the family of Prof. Ocholla, in particular his loving wife, Mrs. Lyudumilla Ocholla, and daughter, Diana Ocholla. When I was hungry they fed me, and when I was bored they entertained me.

I wish also to express my sincere gratitude to the University of Zululand's staff at the Department of Library and Information Science for their hospitality and professional support. In particular, I wish to thank Dr. Daisy Jacobs, Prof. Jerry Le Roux, Dr. Janneke Mostert and Mr. Neil Evans. They were a great source of inspiration to me throughout my sabbatical. I also wish to thank Mr. Shadrack Katuu of the Nelson Mandela Foundation in Johannesburg who availed me with literature that was used in the section entitled 'e-records and information security in an e-environment' in chapter thirteen. Overall, Ms. Julia Mosemann of IGI, who was my contact during the preparation of this manuscript, was an extraordinary inspiration. Often when I was on the brink of giving up, Julia was there, 'knocking' in my mailbox and encouraging me to go on. I am even shy to mention how many times Julia extended the deadline for the completion of this manuscript, save to say that this work would have been completed a year earlier. To Julia, I owe the completion of this work.

The final form of this book was accomplished through the professional and meticulous editorial efforts of Ms. Catherine Ocholla. It was amazing how Catherine was able to detect in the writing of this book, contradictions, repetitions, inconsistencies and improper use of both semantics and syntax.

Last but not least, I am sincerely indebted to my wife (Dorah Lyaka Mutula) and children - Caleb Mwanzo Mutula, Barbara Lynn Mutula, Katie Musungu Mutula and Melody Namuma Mutula - who endured my long stay away from home during my sabbatical engagement. Their understanding and patience were the most critical factor in the successful completion of this book. And finally, to the Almighty God, I say Halleluiah!

Stephen M. Mutula
University of Botswana, Botswana

Section 1
Conceptual Framework and Contextual Setting

This section consists of one chapter, Chapter 1, which sets the scene and unpacks the meaning of the core concepts discussed in this book, namely digital economies, SMEs, and e-readiness. It establishes the nexus between these key concepts, formulates the context, and maps out the rest of the book.

Chapter 1
Introduction

DIGITAL ECONOMY CONCEPT

The 'Digital economy' is sometimes used synonymously with 'information society', which emerged back in the 1960s to describe a futuristic society that is highly dependent on information (Bridges.org, 2001; Computer Systems Policy Projects, 2000). Martin (1997:87) further associates the concept with 'information economics' by defining it as a society in which there is a growing rate in the production, distribution and use of information. The 'Digital economy', as term and concept, has been used in this book in keeping with 'information society' as espoused by Schienstock et al. (1999), who view it from an interdisciplinary perspective to describe:

- An information economy;

DOI: 10.4018/978-1-60566-420-0.ch001

- A post-industrial society;
- The end of the industrial labour society;
- A knowledge society;
- An 'informatized' industrial society; and
- A learning society.

The concept of the information society as a synonym for an information economy was put forward by the anthropologist and biologist Todao Umesao. Later, in 1963, Joho Sangyo Ron wrote about "knowledge industries" and postulated that the "electronic industries", especially information, communication, and education, would be as pivotal for the structural transformation of the industrial society as the intermediary industries - transportation and heavy industries - were for the transition from the agricultural to the industrial society. The notion of an information society as a post-industrial society was proposed by Daniel Bell (1989) when he perceived social change as a multidimensional process, wherein economic sectors, vocational groups, fundamentals of technology, and basic social principles are radically shifted. Consequently, the post-industrial society is founded on the growing importance of the information sector as opposed to the production of goods. Production in such a society is primarily dependent on information rather than on raw materials and energy. Bell further observed that in the industrial society, goods were produced primarily by machines, whereas in the post-industrial society or information society, production is centred on acquiring and utilising information and knowledge through Information and Communication Technologies (ICTs).

The current transformation from the industrial to the knowledge economy is characterised by all sectors and industries of an economy relying largely on information. Moreover, the transformation to a knowledge society is characterised by human work occurring more and more using ICTs. Steinmüller (1981) perceives today's information or knowledge society as an 'informatized' industrial society. Regardless of whether one refers to the knowledge society, the information society or the digital economy, these terms and concepts are all founded on the importance of the computer as a knowledge-based machine that incorporates the knowledge bases and conclusions of experts. Such a society is characterized by an enormous growth in the amount of scientific and non-scientific knowledge; different knowledge-based techniques and types of mechanized knowledge; an expanding information sector; investment of information in technical development or industrial products; and the information explosion (Schienstock et al., 1999). This characterization mirrors Luhmann's view (as cited by Qvortrup, 2007) that the knowledge society is typified by functional systems which are bound to a knowledge base. A knowledge society is therefore characterized by the provision of more products with built-in intelligence and the transformation of organizations into knowledge-based entities. Castells (1997) opines that it is the immediate relationship between knowledge generation and knowledge application through learning that becomes the most fundamental aspect of an information society.

E-READINESS: ORIGIN AND MEANING

The concept of 'e-readiness' emerged because it became necessary to provide a unified framework to evaluate the breadth and depth of the digital divide at macro level between more and less developed countries during the late 90s. E-readiness measures the capacity of nations to participate in the digital economy. It is also perceived as the measure of a country's ability to leverage digital channels for communication, commerce and government in order to further economic and social development (Economist Intelligent Unit/IBM Institute for Business Value, 2008). The inextricable relationship between e-readiness and the digital economy is succinctly captured by Hartman et al. (2000), who state that net readiness is measured as a company's preparedness to exploit the enormous opportunities in the e-economy landscape. According to Grant (1999), a business is "e-ready" when it is able to implement e-business and e-commerce, or as Parker (2000) would put it, it is the preparedness [of a nation, community or an organisation] to operate in an e-business or e-commerce marketplace.

The emergence of the concept of e-readiness attracted the development of various e-readiness assessment tools by different companies, organisations and groups, with each claiming that its tool would help diagnose the current situation and orient steps to narrow the digital divide (Arce and Hopmann, 2002:1). The Computer Systems Policy Project (CSPP) - a public policy advocacy group comprising of IT companies from the US - was the first to use the concept of e-readiness when it developed the e-readiness assessment tool known as the 'Readiness Guide for Living in the Networked World' in 1998. CSPP perceived e-readiness with respect to a community that has (Bridges.org, 2001):

- High-speed access in a competitive market;
- Constant access and application of ICTs in schools, government offices, businesses, healthcare facilities and homes;
- User privacy and online security; and
- Government policies that promote connectedness and the use of the network.

The Centre for International Development at Harvard developed the e-readiness assessment tool known as 'Readiness for the Networked World'. This tool assesses e-readiness according to whether or not a society has the necessary physical infrastructure (high bandwidth, reliability and affordable prices); integrated ICTs throughout businesses, communities and government; and universal access (Bridges.org, 2001). Correspondingly McConnell International, jointly with the World Information Technology and Services Alliance (WITSA), developed the 'Risk E-Business' tool in 2000, which views e-readiness in relation to whether a country has (Bridges.org, 2001): an extensive pervasion of computers in schools, businesses, governments, homes etc; affordable and reliable access to the Internet in a competitive market; free trade; a skilled workforce; high standards of education and training in schools; a culture of creativity; government-business partnerships; transparency and stability in government; an evenly enforced legal system; secure networks; personal privacy; regulations allowing digital signatures and encryption; and consumers' trust in e-commerce security and privacy.

Generally, Bridges.org (2001) views e-readiness from the perspective of those who are not benefiting from IT, and which societies are not 'e-ready.' Many reports generated or referenced by Bridges.org study the digital divide and make recommendations. Like the tools already described, digital divide reports use a range of methods and standards for evaluating e-readiness. However, unlike e-assessment tools, digital divide reports are more concerned with the distribution of technology across society and its impact on the lives of people.

The formal e-readiness assessment tools and surveys described above are complemented by a wide range of other frameworks expressed in digital divide reports, position papers and other kinds of models that can similarly be used for e-readiness assessments. Some of these include:

- The global diffusion of the Internet
- E-commerce readiness assessment
- International survey of e-commerce
- Technology achievement indices

The various definitions and assessment approaches for e-readiness showcase considerable diversity. Agriculture and Food Canada (2001) confirms this notion, finding that there are varying degrees of e-readiness, and each could include any or more than one of the following activities: using e-mail as the most preferred medium of communication; using a website for internal and external communication; selling goods and/or services through the Internet; and making travel arrangements using online Internet services.

From an analysis of the various definitions and perceptions of 'e-readiness' described above, the concept is taken in this book to mean a society characterized by the pervasive availability of access to high speed, good quality networks; the use of ICTs in everyday life, e.g. in business (e-commerce and e-business), government (e-government), education (e-learning), records management (e-records), and content management (i.e. knowledge and information management); high levels of literacy; the presence of national enabling ICT policies; the right physical infrastructure (high bandwidth, reliability, and affordable prices); and business and commercial transactions that are realised via the Internet.

SMEs CONCEPT

SMEs have no standardised definition across different jurisdictions - definitions vary based on employment figures; annual turnover, company capitalisation, fixed assets, etc. The common definition in OECD countries is founded on employment figures; correspondingly, an SME in such countries must have less than 500 employees (OECD, 2002). The South African SME Act defines SMEs as firms that have up to 100-200 employees or a 5 million Rand turnover [US$ 0.5 million] (Gordon, 2003), while in Egypt, the Ministry of Trade's definition of SMEs is based on the number of workers, fixed assets and annual turnover (Maksoud and Youseff, 2003).

The diversity in the various definitions shows that by and large, quantitative parameters, i.e. the number of persons employed in the enterprise, the annual turnover, the level of fixed investment in the enterprise, etc; are used to categorize businesses as SMEs. In other jurisdictions, such as in China, Germany, Japan, Mexico, Taiwan, and South Korea, employment is the main criterion used to determine whether a unit is a SME or not. On occasion, different names are used in the place of SME. In India, for instance, the SME sector is known as the Small Scale Industry. Generally, SMEs tend to be dispersed across an economy or located in clusters, either around a large manufacturing unit or where the raw materials are readily available. More often than not, SMEs also tend to cater for localized niche markets or supply components to the larger units/markets (Tetuja, 2001).

As the business environment grows increasingly saturated with ICTs, the concept of the 'digital SME' has emerged to describe a small or medium-sized enterprise whose products and services are predominantly digital; that mainly uses digital methods to carry out core operations such as marketing, sales and services; and that exploits the benefits of digital methods to a considerable extent (BEEP, 2003). The digital economy is increasingly being defined according to the extent to which a community provides Internet access and connectivity, especially to its business entities, as this would enable them to extend their market reach and partake in global trade. The belief held by Dr. Hamadoun Toure (1996) of the International Telecommunications Union that 1% of access to broadband today equates to 99% in terms of opportunity, is implicit in the potential enterprises (such as SMEs) and other entities can leverage in the digital economy.

CONCLUSION

This chapter has demonstrated the inextricable relationship between the digital economy, e-readiness and SMEs. The digital economy is largely made up of SMEs that rely on digital technologies for electronic business transactions. In order for SMEs to succeed in such an economy, they need a conducive environment that offers intellectual capital, global business reach, and supports continual innovation and quality improvement. Furthermore, they need an environment that has attained a reasonable level of e-readiness with respect to technological infrastructure, policy and legal frameworks, and an enabling business environment. The policy and managerial implications advanced in this chapter require governments to ensure that infrastructure development and enabling policies are in place. These would facilitate an e-ready environment that can accelerate the move towards an inclusive global information society where individuals and businesses alike are empowered to create, receive, share and utilise information for their economic, social, cultural and political development. In terms of management interventions, the inextricable relationship between the digital economy and e-readiness demonstrates the need for various stakeholders in the SME industry, including government, SMEs' owners, the private sector and academia, to work together to create an environment conducive to small business growth in the information age.

ORGANIZATION OF THE REST OF THE BOOK

The whole book is organised in 9 sections containing 19 chapters. Section 1 has already been described.

Section 2: Digital Economy Growth and Infrastructure

This section consists of four chapters – Chapter 2 (Growth of the Digital Economy), Chapters 3 (Digital Economy Components), 4 (SMEs Industry) and 5 (Digital Divide). Generally, the purpose of this section is to introduce the reader to the digital economy environment and its components, including the inadequacies of the components that give rise to the digital divide. Chapter 2 (Growth of the Digital Economy) characterizes the digital economy and covers the history and evolution of the concept 'digital economy', drivers of the digital economy, and benefits of the digital economy. The chapter aims to define the digital environment in which SMEs now operate and the challenges and opportunities of such an environment for businesses, including SMEs. Chapter 3 (Digital Economy Components) defines the major entities that make up the digital economy, including government; policies and regulations; the Internet, World Wide Web and electrical infrastructure; hardware and software industries; telecommunications industry; digital service providers; e-business and e-commerce; information and knowledge management systems; intellectual property rights; telecommunications service providers; human capital and knowledge workers; research and development (R&D); and emerging technologies.

Chapter 4 (SMEs Industry) covers the contribution of SMEs to the national economy, the international status of SMEs (in Asia, America, Europe and Africa); and barriers affecting the SMEs sector, with a case study of the agricultural sector in Africa (technology transfer challenges, intellectual property rights challenges; and micro finance challenges). Chapter 5 addresses issues of the digital divide that arise out of inadequacies in the digital components (Chapter 3) as well as barriers to addressing the challenges in the digital economy environment (discussed in Chapter 4).

Section 3: E-Readiness Assessments

This section consists of Chapters 6, 7 and 8 (E-Readiness and its Assessment, E-Readiness Assessment Methods and Tools, and E-Readiness Assessment of SMEs, respectively). Chapter 6 (E-Readiness and its Assessment) covers the international status of e-readiness across different jurisdictions (Europe, North America, Asia-Pacific, Middle East and Africa), its implications for businesses (especially SMEs), and the importance of e-readiness assessments for businesses and governments. The purpose of Chapter VI is to demonstrate how the level of e-readiness affects the competitiveness of businesses, and also to highlight managerial and policy implications.

Chapter 7 (E-Readiness Assessment Methods) covers measures of e-readiness, macro and micro e-readiness assessment tools (including the tools names, the developers of the tools, intended applications of the tools, the countries in which the tools have been used, and an overall analysis of the tools), strengths and weaknesses of e-readiness assessment tools, and choosing the right assessment tool. The chapter aims to highlight the different types of tools available for use by SMEs or governments that wish to measure their e-readiness and gauge their position in the global digital economy. It also illustrates that while e-readiness tools for SMEs are limited, macro e-readiness tools that are used for measuring national e-readiness may be applied with adaptations.

Chapter 8 (E-Readiness Assessment of SMEs) examines how the e-readiness of SMEs is assessed. It covers the strengths of SMEs over large scale enterprises in the digital economy; the types of transactions that SMEs undertake online; the e-readiness adoption process among SMEs; international e-readiness status of SMEs (in North America Europe, the Middle East, Asia-Pacific, the Caribbean, Africa); and how to get SMEs to reach an e-readiness status. The purpose of this chapter is to enable policy makers and managers of SMEs to compare the relative positions of SMEs in the global business environment so that they may put in place requisite policies and institutional infrastructure to enhance SMEs competitiveness in the digital economy.

Section 4: Digital Economy Applications

This section consists of Chapters 9 and 10 (E-Government and E-Business/E-Commerce, respectively). Section 9 expands on the previous section by examining how e-readiness is critical in the implementation of e-applications such as e-government and e-business/commerce. Governments have a major role to play in providing an environment that is conducive for electronic business transactions by laying requisite infrastructure and enhancing a legal and policy-driven framework. Chapter 9 (E-Government Readiness) therefore covers the nexus between e-government and the digital economy, applications of e-government in society, global e-government trends, global e-government rankings, e-government survey methodologies, benefits of e-government systems, challenges of implementing e-government systems, and the policy and managerial implications of implementing e-government systems. Chapter 10 (E-Commerce and E-Business) covers e-commerce and e-business as major applications of the digital economy; the development of e-commerce; types of e-commerce; determinants of e-commerce adoption by SMEs; e-government/e-commerce nexus; e-commerce implementation in e-government settings; e-commerce readiness (connectivity, business environment, legal and regulatory environment, social/cultural infrastructure, etc); methodologies for assessing e-commerce readiness; and e-commerce challenges. The chapter aims to provide information that can prepare SMEs for an e-commerce environment. It also aims to show how e-commerce may offer SMEs ways of finding new customers and suppliers, especially in international and regional markets which they have not easily been able to reach before.

Section 5: Content, Knowledge and Records Management

This section consists of three chapters: Chapters 11 (Local Content Management), 12 (Knowledge Management) and 13 (E-Records Management and Readiness). Chapter 11 begins by unpacking the concept of content and its derivative, local content. The chapter then discusses the value of local content in enhancing business competitiveness. Also discussed are local content development initiatives at international, regional and national levels, with mention of the agencies that are involved in these initiatives. Challenges of local content development are explained, as are the prospects of local content for SMEs. The chapter also elucidates on policy and managerial roles in local content management and its applications to SMEs. Overall, the chapter aims to demonstrate the value of relying on locally available products and services, in part because they would be cheaper and more easily understood.

Chapter 12 (Knowledge Management) reviews the growing number of studies on knowledge management (KM) as they apply to SMEs. The chapter covers the benefits SMEs can accrue from knowledge management, the meaning and evolution of KM, KM strategies, KM general adoption practices, KM processes and activities, KM technologies, KM challenges, and the policy and managerial responsibilities of KM. The chapter aims to promote awareness among SMEs, governments and SMEs' managers of the value of harnessing knowledge, human capital and know-how for competitive advantage in increasingly competitive electronic business environments. Given that the digital economy is knowledge intensive, SMEs must wake to the fact that without investing in human capital and effectively managing the intellectual capital already in their possession, they may not be able to compete in the digital economy.

Chapter 13 (E-Records Management and Readiness), introduces the concept of (e) records management and observes that electronic records management has not been given much attention by SMEs; this despite an economy where most transactions occur online, meaning that e-records are increasingly being generated and must be meticulously managed. The concept of freedom of information (FOI) is discussed in the context of how it facilitates access to publicly held information, which is necessary for promoting business activities within SMEs. The chapter also covers the role of e-records management; the nexus between records management, SMEs and FOI; global initiatives for enhancing FOI; e-records readiness; e-records generation and use; e-records management; e-records management models; e-records and information security; e-records authentication - digital signatures, digital certificates, etc; e-records management challenges; and policy and managerial implications of records management for SMEs. The chapter's intention is to showcase the importance of records management in general and e-records management in particular in business environments. Without effective records management, businesses cannot leverage their intellectual capital to effectively compete in the digital age. Poor records management, especially in governments, also erodes business confidence and discourages firms from investing in such jurisdictions because of delays experienced either in setting up businesses or litigation resulting from poor or a complete lack of documentation or documented procedures.

Section 6: Information Needs and Capacity Building in SMEs

This section consists of Chapters 14 and 15 (Information Needs and Access in SMEs; and Capacity Building in SMEs, respectively). Section 6 serves two main purposes, namely to demonstrate the difficulties that SMEs face when searching for information to meet their business needs, and to try and show how capacity building could empower SMEs that are able to recognise their information needs, identify relevant sources of information, and apply such information.

Chapter 14 (Information Needs and Access in SMEs) highlights why policy developers and managers should find it important to understand the information needs of SMEs. This chapter therefore covers information management in SMEs; information services and access in SMEs; information use by SMEs; factors affecting information seeking in SMEs; challenges of information seeking in SMEs; and policy and managerial roles in meeting the information needs of SMEs. The chapter's aim is to show how SMEs can effectively identify and gain access to critical business information (markets, sources of venture capital, legal advice, use of technology, best practices, policies and regulations, etc)

Chapter 15 (Capacity Building in SMEs) maintains that one of the main challenges facing SMEs is skills shortages; consequently, it is urgently necessary to make capacity building a real priority. This chapter covers SMEs' capacity building initiatives; education and training; case studies/best practices in capacity building – Europe, North and South America, Asia-Pacific, The Caribbean, and Africa; information literacy needs of SMEs' employees; and the policy and managerial roles in capacity building for SMEs.

Section 7: Globalisation, Trends and Best Practices in the Digital Economy

This section consists of two chapters: Chapters 16 and 17 (Globalisation of the Digital Economy; and Trends and Practices in the Digital Economy, respectively). Chapter 16 (Globalisation of the digital economy) begins with the ICT-driven impact of globalisation on commerce and the competitiveness of SMEs. This is followed by a discussion of global business competitiveness by region - North America, Western Europe, Eastern Europe, Asia-Pacific, Latin America and the Caribbean, the Middle East, and Africa. The opportunities the digital economy engenders for SMEs are then followed by the role policy makers and managers of SMEs can play in this respect. The chapter aims to provide best practices from around the world from which SMEs, especially those in the developing world, can learn.

Chapter 17 (Trends and Best Practices in a Digital Economy) expands on the previous chapter by focusing on the global initiatives of governments and private agencies to enhance international trade, especially with regard to e-commerce. The initiatives are categorised by continent - North America (led by the United States and Canada); Europe (spearheaded by the European Union and individual countries such as the UK and Germany); Nordic countries (Norway, Denmark and Finland); and Asia (South Korea, Mongolia, Taiwan, New Zealand, Hong Kong, China, India and Japan). Africa does not yet offer much in terms of examples of best practices in the digital environment. The chapter moves on to discuss the concept of a learning organisation in the digital economy.

Section 8: Challenges of the Digital Economy, SMEs and E-Readiness

This section consists of one chapter - Chapter 18 (Challenges of the Digital Economy). The Chapter collates the challenges discussed thus far in each chapter of the book. The challenges are categorised by type as follows: internet-related challenges; e-commerce challenges; political and jurisdictional challenges; policy, regulatory and legal framework challenges; and challenges peculiar to SMEs (skills shortages, lack of awareness, online challenges, competition, consumer protection, taxation, trade investment barriers, etc). The chapter aims to showcase the challenges SMEs are likely to face and offer suggestions on how these may be ameliorated.

Section 9: Conclusion, Policy and Managerial Recommendations

This is the final section of the book and consists of one chapter - Chapter 19

Conclusion and the Way Forward

The chapter summarises the main issues discussed in the book and proffers policy and managerial recommendations that can empower SMEs to compete effectively in the global digital economy.

REFERENCES

Agriculture and Food Canada. (2001). *The e-readiness of Canadian agriculture and food Canada SMEs*. Retrieved on March 21, 2004, from http://www.agr.gc.ca/misb/fb/food/ecommerce/eready/introduction_e.html

Arce, M. E., & Hopmann, C. (2002). *The concept of e-readiness and its application in developing countries: Methodological problems and results for the e-readiness of Nicaragua* (pp. 1-25). Managua: Nicaragua National Chamber of Industries.

The BEEP Knowledge System. (2003). *The digital SME*. Brussels: EC.

Bell, D. (1989). *Die Nochindustrielle Gesellschaft.* New York: Frankfurt am Main.

Bridges.org. (2001). *Comparison of e-readiness assessment models: Final draft.* Retrieved on July 16, 2003, from http://www.bridges.org/eredainess/tools.html

Castells, M. (1997). *The rise of the network society.* Oxford: Blackwell Publishers.

Computer Systems Policy Projects. (2000). *The CSPP readiness guide for living in the networked world.* Retrieved on August 12, 2004, from http://www.cspp.org/projects/readiness/

Economist Intelligent Unit / The IBM Institute for Business Value. (2008). *E-readiness rankings 2008*. London: EIU. Retrieved on October 25, 2008, from http://a330.g.akamai.net/7/330/25828/20080331202303/graphics.eiu.com/upload/ibm_ereadiness_2008.pdf

Gordon, G. (2003, February 16). SME survey. *Sunday Times Business Times, 15.*

Grant, S. (1999, September). E-commerce for small businesses. *IEC'99 Conference.*

Hartman, A., Sifonis, J. G., & Kador, J. (2000). *Net ready: Strategies for success in the e-economy*. USA: McGraw-Hill.

Maksoud, S. S., & Youseff, M. A. A. (2003). *Information and communication technology for small and medium enterprises in Egypt* (pp. 1-15). Cairo: SME Development Unit-Ministry of Foreign Trade.

Martin, W. J. (1997). *The global information society*. Aldershot: Aslib Gower.

OECD. (2002). *Measuring the information economy*. Retrieved on January 21, 2004, from http://www.oecd.org/document/5/0,2340,en_2649_34449_2765701_1_1_1_1,00.html

Parker, S. (2000, April). A survey of small business in Colorado. *CCCOES*.

Qvortrup, L. (2007). The public library: From information access to knowledge management: A theory of knowledge and knowledge categories. *IR Information Research, 12*(4). Retrieved on October 25, 2008, from http://informationr.net/ir/12-4/colis/colis17.html

Schienstock, V. G., Bechmann, G., & Frederichs, G. (1999). Information society, work, and the generation of new forms of social exclusion (SOWING)-the theoretical approach. *TA-Datenbank-Nachrichten, 1*(8), 3-49. Retrieved on August 5, 2008, from http://www.itas.fzk.de/deu/tadn/tadn991/scua99a.htm

Steinmüller, W. (1981). Die zweite industrielle revolution hat eben begonnen. *Kursbuch, 66*.

Tuteja, S. K. (2000, September 28-29). Small business survival & growth in the context of globalization. The Indian scenario. *Paper presented at the APO Study Meeting on Small Business Survival & Growth*, Jakarta, Indonesia.

Toure, M. (1996, September 4-6). Commercialization and transfer of agricultural technologies in sub-Saharan Africa. *A roundtable workshop University of Maryland Eastern Shore Princess Anne, Maryland.*

Section 2
Digital Economy Growth and Infrastructure

This section consists of four chapters: Chapter 2 (Growth of the Digital Economy), Chapter 3 (Digital Economy Components), Chapter 4 (SMEs Industry), and Chapter 5 (Digital Divide and SMEs). Generally, the purpose of this section is to introduce the reader to the digital economy growth, environment, and its components, including the inadequacies of the components that give rise to the digital divide. Chapter 2 characterizes the digital economy and covers the history and evolution of the concept 'digital economy,' drivers of the digital economy, and benefits of the digital economy. The chapter aims to define the digital environment in which SMEs now operate and the challenges and opportunities of such an environment for businesses, including SMEs. Chapter 3 defines the major entities that make up the digital economy, including government, policies and regulations, the Internet, World Wide Web, and electrical infrastructure, hardware and software industries, telecommunications industry, digital service providers, e-business and e-commerce, information and knowledge management systems, intellectual property rights, telecommunications service providers, human capital and knowledge workers, research and development (R&D), and emerging technologies.

Chapter 4 covers the contribution of SMEs to the national economy, the international status of SMEs (in Asia, America, Europe, and Africa), and barriers affecting the SMEs sector, with a case study of the agricultural sector in Africa (technology transfer challenges, intellectual property rights challenges, and micro finance challenges). Chapter 5 addresses issues of the digital divide that arise out of inadequacies in the digital components (Chapter 3), as well as barriers to addressing the challenges in the digital economy environment (discussed in Chapter 4).

Chapter 2
Growth of the Digital Economy

EVOLUTION OF THE DIGITAL ECONOMY

As pointed out in the previous chapter, the concepts 'digital economy', 'information economy', 'information society' and 'knowledge society/economy' are inextricably intertwined and are often used interchangeably. However, the concept 'information society' is increasingly being seen as a unifying term. In defence of why this is so, Schienstock et al. (1999) have argued that the notion of an 'information society' is presented as a strategic aim meant to overcome current social stagnation. The central argument is that the information society would create and secure millions of new jobs; guarantee economic competitiveness; bring ecological advantages; intensify democracy; and revolutionize our ways of living and working with the help of new communication technologies.

DOI: 10.4018/978-1-60566-420-0.ch002

In 1962, Machlup was instrumental in studying information society in the United States. He coined the concept 'knowledge industry' and made significant contributions to information economy research. Martin (1997) defines an information society as a society characterized by rapid growth in the use of information - a society in which quality of life and economic development depend largely on information and its exploitation, the increasing use of computers, the commoditisation of information, e-commerce, and the use of technology for community development. To Deutsch (1983), information societies are national economies in which more than half of the workforce is employed primarily in information-oriented occupations, and in which the net product of these activities is greater than half of the Gross National Product (GNP). In the OECD countries, all gainfully employed individuals who produce, process and distribute information, or maintain the apparatus for its preparation and mediation, are information workers.

The belief that knowledge in its different forms is the engine of economic progress is growing; essentially, an economic system not based on knowledge is not possible. This is because in the current economic climate, a firm's access to know-how is perceived to be the main determinant of its growth. In the UK for instance, a knowledge-driven economy is considered to be one in which the generation and exploitation of knowledge play the predominant role in the creation of wealth (Department of Trade and Industry, 1999. In such an economy, everyone has access to large amounts of information anywhere, anytime and in various requisite formats. Moreover, in such information economies, knowledge is perceived to be the most important determinant of the standard of living, more even than land and labour. Unlike the industrial economy where machines dominated productivity, in today's economy it is information that defines society, with intellectual content the dominant source of value across jurisdictions.

In an information economy, the processing and exchange of information and knowledge are at the core of its activities. The US economy, for example, has been performing well because it has been driven by revolutions in Information Technology (IT), which have affected growth, employment, inflation, and productivity. Underlying factors propelling growth in this new economy include technological innovation, e-commerce and digital transformation, higher education and IT skills, open trade, and a balanced budget. The vast amount of information made accessible via the computer caters for the process of knowledge accumulation by networking all sources of knowledge, thereby facilitating the globalization of knowledge (The New Economy Task Force, 1999; Margherio, 1998). As in the US, in Europe the growth of the information service sector is largely attributable to the region's industrial transformation.

CHARACTERISTICS OF DIGITAL ECONOMY

The digital economy is characterised by numerous phenomena in the global business environment, which include but are not limited to globalisation, digitisation, virtualisation, and disintermediation. The phenomenon that is globalization has internationalised the economy by expanding market reach, scope and speed. Boundaries have given way to electronic borders that allow millions of businesses to carry out online transactions in a seamless fashion. Likewise, digitization has removed the need for a human interface, meaning that goods and services no longer need to be tangible, leading to the development of e-books, digital music and the like. Other than this, virtualization has made physical location and proximity irrelevant, with business transactions able to take place in a virtual environment involving many to many or one to many communications. Finally, disintermediation has made it possible for consumers to bypass middlemen, retailers and other intermediaries and directly reach producers.

The digital economy applications like e-commerce are largely driven by information and knowledge resources facilitated by widespread access to computers and the Internet. Nicholas and Rowlands (2000) developed a model for the progression of data into information, information into knowledge and knowledge into wisdom. Data occupies the basic level of the model and consists of raw facts or observations upon which no meaningful decision can be made until these have been processed into information. According to the model, once data has been subject to refinement, it evolves into information, which can ideally be interpreted such that meaningful decisions can be derived from it. The third stage of the model is knowledge, a value-added derivative of information that is relevant and actionable. Following this, is the final stage in the Nicholas-Rowlands model, i.e. wisdom, which by implication is the ability to perceive or determine what is good, true and sound.

The digital economy has information and knowledge as its main commodities for production and consumption. Consequently, sound information and knowledge management (KM) are critical for organisations wishing to improve their performance. Information management centres on a number of different tasks, including the creation and maintenance of meta-information, searching for documents and other data objects, and viewing and retrieving information (Information Management Consultants, 2003). Further tasks include information uploads; data extraction, publication and distribution; and providing access to information. Like information management, effective knowledge management (KM) is critical for the growth of the digital economy. From the viewpoint of KM, human ingenuity is the most important feature in the emerging digital economy. Constant technological innovation makes success in the marketplace increasingly dependant on the skills of the workforce.

In the digital economy, nothing is more precious than human ingenuity, and the promise of the digital economy therefore lies in this new type of worker.

KM is broadly concerned with the collection of organisational practices related to generating, capturing and disseminating skills and know-how. It is mainly centred on re-using knowledge and skills in order to enable an organisation to achieve its goals. KM processes are largely dependent on technology-based tools that facilitate knowledge production, reconstruction, interpretation, contextualization and application. Although KM is said to have evolved in corporate environments, spearheaded mostly by the revolution in ICTs, knowledge management ideas and concepts have been around for millennia and Alexander the Great, Caesar, Napoleon and Henry Ford are credited as pioneers of knowledge practices. However, KM only started gaining momentum in organisations in the mid 1980s following rapid technological development and the gradual awareness that information and knowledge are strategic resources.

Knowledge management in the evolving digital economy has increasingly become an important theme in public organizations, although private firms have long considered it to be a critical determinant of a firm's competitiveness. In the private sector, the importance of KM was recognized more than a decade ago, and the sector now boasts fairly advanced knowledge management systems. In Europe, KM ranks high on the management agenda of most central government organisations across OECD member countries; increasingly, strategies of managing knowledge are being devised, and knowledge management is clearly signalled as a top priority in business (Organisation for Economic Cooperation and Development, 2003).

The motivations for integrating KM in both private and public organisations in the digital economy are worth noting. For one, the pressures brought on by competition and incentives to lower costs are increasing. Moreover, as modern organisations become more knowledge-intensive, such organisations cannot be expected to function properly if they do not have good mechanisms with which to share knowledge across their different units in order to enhance policy-making and service delivery. The public sector also operates in an environment where, for reasons of greater public interest, transparency is widely encouraged and the bulk of knowledge is widely accessible.

The digital economy is essentially a knowledge-intensive economy where goods and services are increasingly being produced in the form of intangible capital, making knowledge an important measure of the competitiveness of an organisation. Organisations increasingly compete with each other for the use and provision of knowledge-intensive products, such as human capital and new technologies. The importance of KM in organisations is now increasingly acknowledged when it is realized that pressures of the knowledge economy on organisations compel them to consider knowledge management as a way to enhance productivity and efficiency. A study of knowledge management practices in the public sector in OECD's countries found that one of the main motivations centres on concerns around efficiency and productivity (Organisation of Economic Cooperation and Development, 2003). The knowledge economy in which governments and entire societies now find themselves is founded on a structural transformation in which the rapid creation of new knowledge, and the improvement of access to knowledge bases, is factors that are believed to increase efficiency, innovation, the quality of goods and services, and equity.

The digital economy is also characterized by economic activities that are now possible without the physical movement or involvement of people, things, money, etc; rapid developments in the globalization of economic activities; the transfer of value and accumulation of assets via digitized networks; ICT's current situation as the foundation of the digital economy; e-commerce and digital information's encroachment into all aspects of people's lives; the 'dumping' of previous (economic) rules in favour of new ones more suited to the digital economy; and dramatic increases in business opportunities for SMEs through the effective application of IT (Ministry of International Trade and Industry-Japan, 1997).

The new economy, as the digital economy is known in the US, is considered to be a knowledge and idea-based economy where the key to higher standards of living and job creation is the extent to which innovative ideas and technologies are embedded in services, products, and manufacturing processes. Moreover, it is an economy where risk, uncertainty, and constant change are the rule rather than the exception - an economy where hierarchical organizations are being replaced by networked learning organizations, and where there is a progressive drive for increased productivity and higher incomes, more knowledge-based jobs, an expanded number of stakeholders, and greater access to information by citizens. Comparatively, in the old economy information was a valuable commodity, often only available to those who had the means to pay for it. Now with the Internet economy, information is and continues to become more ubiquitous (The New Economy Task Force, 2000).

DRIVERS OF THE DIGITAL ECONOMY

Various factors have given impetus to the transformation from the industrial economy to the emerging digital economy. Within European Union member states, building trust and confidence in e-commerce transactions is an important consideration. It goes without saying that consumers and businesses should be confident that the seller and buyer are who they say they are and that transaction mechanisms are available, legal and secure (The European Commission, 2005). The European Commission found that one of the greatest barriers faced by SMEs wishing to start online businesses was the lack of trust and confidence in terms of the legal implications, especially when trading across jurisdictions. Driving the digital economy also depends on removing obstacles that hinder transactions in an electronic environment. Such obstacles include legal uncertainties with regard to the validity and enforcement of electronic contracts, protection of privacy and personal data, reliability of payments, recourse for errors and fraud, and security and copyright issues, to name a few. A favourable regulatory framework that accommodates the electronic environment is important because it can enhance: the growth of SMEs; higher skills profiles; a sustainable and rich pool of graduate labour; Internet growth and usage; wide broadband coverage; and self-regulation, especially with respect to the use of the Internet.

The digital economy can also be enhanced by the extensive deployment of broadband communication to businesses, government facilities and residences. Further growth enhancements would require enabling government regulations and incentives; effective knowledge workers, education and training; the deployment of government and private-sector programs that promote digital democracy and ensure that all sectors of society benefit from the broadband revolution; and expanding citizen participation in government decision-making processes through e-governance. The new economy would also be accelerated if attention was given to the propagation of electronic commerce; computerization of the public sector; improving information literacy; and installing advanced network infrastructure. Ohshima (2000), in the context of Japan, lists the changes in the business environment that could accelerate the evolution of the digital economy as follows: diversifying the needs of consumers; increasing choices by which consumers can access goods and services; enhancing opportunities for consumers to access cheaper commodities more easily; increasing demand by customers for stronger competitive prices; improving quality; and shortening lead-time.

The advances in ICTs have occasioned pressure for the digitization and reform of both internal activities and inter-activities concerned with customers and related trading partners. As a result of these changes, businesses now need to adapt to the network society and install business processes based on information and communication technologies for both front and back offices. Measures that would enhance e-business include: an enabling legislation system; diffusing electronic payments for businesses; lowering communication costs; and developing the skills necessary for using e-commerce business models. Moreover, because it is not easy for a single SME to acquire enough human resources, know-how, technology, and facilities, SMEs can use their characteristic mobility and flexibility and collaborate with others. They can also construct mechanisms to execute collective business activities over the Internet.

Since the 90's, the revolution in computer technology has had a profound impact on the digital economy. In the early 90s, PCs started to penetrate and pervade homes and businesses - spurred on by falling IT prices - making e-commerce possible. OECD (2000) suggest that two major factors likely to influence the future expansion of e-commerce are the extent to which IT companies invest in network capacity and the speed of data transmission. Investment in ICTs is also believed to be one of the main drivers underpinning the remarkable performance of the US economy. As Martin (1997) noted, the generational changes in computing that increased computation's power by a factor of ten every five to seven years produced gains for the entire information technology revolution. This development no doubt affected the evolution and revolution of the Internet and the World Wide Web. In turn, the evolution of the Net has provided further impetus to the growth of the digital economy. Part of why the Net has evolved rapidly and catalysed the evolution of the digital economy is because it is not tightly controlled by any major public or private bureaucracy.

The emergence of the information economy has also been attributed to a combination of other factors, including acknowledgement of the unrealistic assumption behind the neoclassical theory, a reaction to the failures of government and business policies, and the potential contribution of intelligent electronics to enhanced communication, computation and control (Martin, 1997). The digital economy may also be perceived as an evolution of capitalism, reflected in the process of technological advancement and the 'informatization' of firms in the manufacturing and service industries, especially in the US and Europe.

The digital economy's continuous evolution may further stem from a need to look beyond the factory system as the only possible method of increasing material wealth. Barbrook (n.d) noted that the rapid spread of personal computing, and now the Net, are the technological expressions of the desire of many people to escape from the petty controls of the shop-floor and office. The digital economy has grown out of a need for novelty, occasioning the consumer society to start looking for something that exists beyond money. This may explain why from the 1960s onwards, many workers started seeking more autonomy in their jobs and more freedom in their personal lives by abandoning traditional conservatism. It is also possible that the digital economy was given impetus when wages started falling in the US and the levels of unemployment started rising in EU member states.

The evolution of the digital economy may also be attributed to the natural evolution of societies as they develop steadily across generations. This development rests on dramatic socio-economic changes and fundamental transformations in the traditional economy, as was exemplified in the replacement of the agrarian society with the industrial society. During the 1970s and 1980s, there was also a prolific output of general publications which may have pointed to the evolving of the digital economy. During that period, the world appeared to be undergoing some form of a new industrial revolution or entering a new phase, referred to as the 'post-industrial society' (Bell, 1989), the 'information economy' (Porat, 1976) or the 'third wave' (Toffler, 1980).

The emerging information or digital economy in the US has been attributed to the following (The US National Science Foundation et al., 2005):

- Globalization of communications and commerce
- Commoditization of ICTs and codified knowledge
- The growing role of scientific research in innovation
- Advanced, integrative information infrastructure
- Modularization, vertical desegregation and outsourcing
- Expanded value chains and clusters with new categories of actors.

As can be seen, the drivers of the information economy are many, and it is difficult to pigeonhole one that has had the most influence. However, it can safely be stated that since the mid 1990s, the Internet has significantly influenced the selling of products of small-sized enterprises via virtual shops, thus contributing to the general development of an Internet-based economy, especially in developed countries such as the USA, Japan and Canada. The rise in the digital economy is reflected in the commerce statistics in the United States for the years leading up to 2000, which showed that the percentage of business-to-consumer (B2C) transactions, in terms of turnover capacity, equated to about 80% of the total e-commerce in the world. The same statistics also showed that the remaining 20% belonged to business-to-business (B2B) e-commerce (Cabbar, 2002).

The digital economy has also been motivated by the need for companies to increase productivity, cut costs and enhance customer services; increase choices; access more information when making purchasing decisions; save time; and make shopping more convenient for consumers. Businesses are also motivated to invest in ICT and electronic commerce because the lower operating costs of many Internet businesses may, in turn, drive to reductions in prices or improvements in quality. Significantly, consumers in the digital economy are no longer limited by shopping in stores within a reasonable driving or walking distance. Instead, via online platforms, customers can shop at stores in any place in the world. Online support tools, order statuses, product availability and pricing, and technical support, are increasingly becoming available to customers.

A lot more can be done to accelerate the transformation process. The New Economy Task Force (2000) suggests that the move can be accelerated if: there is an increase in federal support for research; federal tax code can be made an agent of innovations in the new economy; companies can be encouraged to invest more in training; and government and universities can work with industry to boost individuals with high-tech skills. Moreover, providing security for e-commerce to prevent and stop abuse while avoiding over-regulation; developing a digital government; investing in requisite infrastructure in all regions; and supporting public-private partnerships to empower community organizations through information technologies, can catalyse the growth of the digital economy.

The digital economy demands higher levels of education from those entering the job market, more high-tech skills, and opportunities for lifelong learning for workers to keep up with rapid developments in technology, globalization and new business practices. Within the digital economy, a more educated workforce is critical not only to raise per-capita incomes, but also to reduce income inequality. The New Economy Task Force (2000), in the context of the United States, noted that in the old economy, it was assumed that workers with basic skills would be hired by companies that would then train them in company-specific skills, following which the workers would advance up the career ladder. Comparatively, in the new economy, requisite upfront skills are critical. Moreover, companies cannot be relied on to offer their workers with training because of increased competitive pressures coupled with reduced employment tenure.

The growth of e-commerce has benefited greatly from the open structure of the Internet. The low cost of using the Internet permits the interconnection of new and existing information and communication technologies, and offers businesses and consumers a new and powerful information system and form of communication. This makes it possible for buyers and sellers to come together in more efficient ways, creating new marketplaces and opportunities for the reorganisation of economic processes. As a distribution channel, the Internet has also changed the way products are customised, distributed and exchanged, and how businesses and consumers search and consume products. Coppel (2000) notes that in countries such as Australia, the United Kingdom and the United States, governments are beginning to reorganise the management of public procurement systems - equivalent to some 10% of GDP - over the Internet, opening the prospect of sizeable B2G transactions. The technology is also being used by governments for the transmission or receipt of information (G2B, G2C) to improve the convenience and lower the cost of payment systems and tax compliance (C2G), and by businesses to manage after-sales service and develop direct consumer marketing.

The increasing share of e-commerce taking place between businesses can be linked to the rapid migration of supply chain management from relatively expensive closed EDI networks to the Internet. B2B e-commerce is also being driven by the potential for businesses to obviate intermediaries and deal directly with suppliers, thereby lowering purchasing and inventory costs, and the ability to use the technology to promote a more efficient and effective customer service. Electronic Data Interchange (EDI) was developed earlier as a standard for processing and transmitting information between computers over private communication networks, also known as value-added networks. It requires expensive and complex customized software, dedicated communication links and in many cases, strictly compatible equipment. The main users of the service are large businesses and their first-tier suppliers. The EDI standard is now less used because of its relatively high costs when compared to the more flexible Transmission Control Protocol/Internet Protocol (TCP/IP) based systems (Coppel, 2000).

Consumers' profile has also contributed significantly to the evolvement of the digital economy. Income, education and age are the main factors determining the profile of Internet users and B2C e-commerce buyers. For instance, in the United States and OECD countries, the rate of Internet access and use among university graduates is about three times higher than that of students or graduates with a high school diploma or less, and over half the population with annual household incomes above $50 000 access the Internet compared to less than 20% of those with annual household incomes of $20 000 or less (OECD, 2005a). Moreover, the business use of the Internet reveals a significant dichotomy, with usage and Internet penetration substantially higher among large firms than in small-sized enterprises, and also differing widely according to the sector in which the firms are engaged. OECD further notes that the rapid growth in the number of people who use the Internet and its commercial applications has been stimulated by technological innovations and their diffusion. Together with economic and regulatory reforms, notably in the telecommunications sector, these technological advances have lowered the cost and improved the quality of Internet access. Cheaper computers and telephone connectivity have additionally stimulated their diffusion into businesses, while usage costs have dropped, in part linked to the liberalisation of fixed telecommunications networks.

The growth of the digital economy is bound to accelerate in the coming years across all sectors of society as the number of people connected to the Internet multiplies and as its commercial use grows. The growth will be driven by four types of economic activity: the use of the Internet, driven by dramatic increases in computers, software services and communications investments; electronic commerce among businesses, with productivity improvements in the use of electronic networks to create, buy, distribute, sell, and offer products and services; the digital delivery of goods and services, such as software programs, music etc; and the retail sale of tangible goods and services that are produced, stored and physically delivered, e.g. computers, software, cars, books, flowers etc. (Meeker and Pearson, 1997).

BENEFITS OF THE DIGITAL ECONOMY

The digital economy, which is largely defined by electronic commerce applications, is widely expected to improve efficiency because of reduced transaction and search costs, increased competition, and more streamlined business processes. Coppel (2000) posits that greater efficiency will manifest itself in lower prices, more frequent price modifications, and a narrower dispersion of prices for identical products. Lower search costs may also possibly lead to Internet consumers being more sensitive to price changes. To Brynjolfsson and Smith (2000), B2C is expected to raise competitive pressure and improve economic efficiency. Part of the reason for this is that certain reductions in cost are offset by higher overheads elsewhere. For example, e-commerce retailers may have a better idea of their clients' preferences, which makes more direct marketing and mass customisation of products possible, and can also lead to the more finely differentiated and sophisticated price discrimination of products. The greatest possibilities for e-commerce to reduce prices exist in goods and services that can be digitised, thereby allowing substantial economies of scale in production and delivery costs (Sachs, 2000).

It is also expected that the development of the Internet and e-commerce could also alter how payments are made for products and services, and this in turn may have implications for the setting and operation of monetary policy. The characteristics of the digital economy may change the extent to which e-commerce facilitates more efficient stock management, leading to lower inventories as a ratio of sales. In addition, increased price competition in product markets may allow the economy to sustain more jobs without raising inflation over time, and might also put greater pressure on companies to curb wage growth and modify the process that generates inflation.

CHALLENGES OF THE DIGITAL ECONOMY

There are several challenges that face businesses in the digital economy environment. These challenges include the small number of skilled workers in jobs that use ICTs; dependence of most businesses on low knowledge-intensity jobs; the bulk of lower skill sector posts that are not suitable for the information economy; wide and growing social and economic divides; and the challenges of fully migrating services online so that they can be availed to all. Scottish Executive (2007) underscores the following challenges of the digital economy:

- The problems associated with low computer literacy
- Slow broadband uptake
- Lower proportion of people in 'e-workable' jobs
- Social and economic inequalities.

CONCLUSION

This chapter expanded on the meaning of the digital economy, which was briefly introduced in chapter one, and the drivers that have given impetus to the evolvement of this new economy. The basic premise of the chapter is that the rapid evolvement of the digital economy is driven largely by the ability of the Internet to connect people to people, people to businesses and people to government, thus creating a highly networked world. The information explosion and the growing recognition that information and knowledge are the key components of a modern economy have also given great impetus to the development and growth of the digital economy. The digital economy has the potential to enable business enterprises, particularly SMEs or any other enterprises that have previously been excluded from mainstream global trade, to compete in the international market.

REFERENCES

Bell, D. (1989). *Die Nochindustrielle Gesellschaft*. New York: Frankfurt am Main.

Brynjolfsson, E., and M. D. Smith (2000). Frictionless commerce? A comparison of Internet and conventional retailers. Management Science, 46(4), 563–586. doi:10.1287/mnsc.46.4.563.12061

Cabbar, Y. (2002, May 2). Enterpreneurship and SME development. *Opening Statement at Black Sea Economic Cooperation Workshop on Internet, E-Commerce, and SMEs*, Ankara, Turkey.

Coppel, J. (2000). E-Commerce: Impacts and policy challenges. *Economics Department Working Papers no.252*. Retrieved on July 20, 2007, from http://www.oecd.org/eco/eco

Deutsch, K. (1983). Soziale und politische aspekte der informationsgesellschaft. In P. Sonntag (Ed.), *Die zukunft der informationsgesellschaft*. Frankfurt am Main.

economy. *Conference for Ministers Responsible for SMEs and Industry Ministers,* Bologna, Italy. Retrieved on August 8, 2004, from http://www.conferenzabologna.ipi.it/Eng/dopo_la_conferenza/documenti/WK3_eng.pdf

European Commission. (2005, November 24-25). Transforming public services. *Report of the Ministerial E-Government Conference,* Manchester, UK. Retrieved on May 10, 2007, from, http://archive.cabinetoffice.gov.uk/e-government/conference/

http://www.oecd.org/LongAbstract/0,2546,en_2649_37461_33779951_70666_119684_1_37461,00.html

Information Management Consultants. (2003). *What is information management?* Retrieved on January 30, 2004, from http://www.willpower.demon.co.uk/infoman.htm

Machlup, F. (1962). *The production and distribution of knowledge in the United States.* Princeton: Princeton University Press.

Margherio, L. (1998). *The emerging digital economy.* Retrieved on July 17, 2008, from http://www.technology.gov/digeconomy/ederept.pdf

Martin, W. J. (1997). *The global information society.* Aldershot: Aslib Gower.

Meeker, M., and S. Pearson (1997). Morgan Stanley U.S. investment research: Internet retail. Morgan Stanley, 2(2), 2–6.

Ministry of International Trade and Industry–Japan. (1997). *Towards the age of the digital economy--for rapid progress in the Japanese economy and world economic growth in the 21st century.* Tokyo: Ministry of International Trade and Industry Japan.

Nicholas, D., & Rowlands, I. (2000). *The Internet: Its impact and evaluation.* London: Aslib.

OECD. (2000, June 14-15). Realising the potential of electronic commerce for SMEs in the global

OECD. (2003). *The impact of laws and regulations on micro and small enterprises in Niger and Swaziland.* Retrieved on March 5, 2005, from

OECD. (2005a). *Communications outlook.* Paris: OECD.

Ohshima, M. (2000). *Local EC business community for SMEs*. Tokyo: Electronic Commerce Promotion Council of Japan.

Porat, M. U. (1976). *The information economy*. Stanford, CA: Stanford University.

Sachs, G. (2000, May 9). *Technology: Internet-commerce*. Washington, D.C.: Global Equity Research.

Schienstock, V. G., Bechmann, G., & Frederichs, G. (1999). Information society, work, and the generation of new forms of social exclusion (SOWING)-the theoretical approach. *TA-Datenbank-Nachrichten, 1*(8), 3-49. Retrieved on August 5, 2008, from http://www.itas.fzk.de/deu/tadn/tadn991/scua99a.htm

Scottish Executive. (2007, February 25). Scoping the impacts on travel behaviour in Scotland of e-working and other ICTs. *Research.*

The New Economy Task Force. (1999). *Rules of the road: Governing principles for the new economy: New economy task force report*. Retrieved on July 18, 2008, from http://www.ppionline.org/ppi_ci.cfm?contentid=1268&knlgAreaID=128&subsecid=174

The New Economy Task Force. (2000*). Making the new economy grow an action agenda*. Retrieved on July 21, 2008, http://www.ppionline.org/ppi_ci.cfm?contentid=1490&subsecid=123&knlgArea

Toffler, A. (1980). *The third wave*. New York: Bantam Books.

U.S. National Science Foundation. Organisation for Economic Cooperation and Development, Research Directorate-General, European Commission, Information Society Directorate-General, European Commission, and U.S. Interagency Working Group on IT R&D University of Michigan. (2005). *Advancing knowledge and the knowledge economy*. Retrieved on July 21, 2008, from http://advancingknowledge.com

Chapter 3
Digital Economy Components

INTRODUCTION

The digital economy consists of various components, key among which include government; policy and regulation; internet, the world wide web (WWW) and electricity infrastructure; telecommunication industry; digital service providers; e-business and e-commerce industry; information and knowledge management systems; intellectual property rights; human capital and knowledge workers; research and development; and emerging technologies.

DOI: 10.4018/978-1-60566-420-0.ch003

Government

Governments are an important component of the digital economy by virtue of their traditional role in providing primary funding for a country's communications infrastructure. They also have an important role to play in sustaining infrastructure development and improving e-readiness. Through national ICT policies, governments provide a national vision for infrastructure development that is aimed at enhancing digital services within their jurisdictions and beyond. A progressive government in the digital economy supplies businesses, citizens and organisations with a clear roadmap for the adoption of technology. A government's investment in digital processes also helps to improve its own operations. For example, governments as early adopters of digital practices provide leadership that other organisations and individuals can emulate. They further create demand for technology and digitally enabled services. A government can also leverage the benefits of ICT by, among other actions:

- Investing in a number of ICT related projects within the public sector to improve public service administration
- Developing incentives to attract investments in the ICT sector
- Maintaining a duty free policy on computer hardware to increase usage
- Engaging in a telecom sector reform project to introduce competition and encourage investment.

Policy and Regulation

ICT or telecommunication policies are fundamental in the digital economy. A conducive business environment is necessary for firms to thrive and benefit from ICTs. This requires a transparent, open and competitive business framework; clear, independent rules of law that are applicable to all firms; mechanisms for the easy set up and dissolution of businesses; transparent, simple and accessible corporate regulation; and equal and stable legal treatment for national and cross-border transactions (OECD, 2004). In order for SMEs to transcend online trust barriers and be able to conduct e-business on any scale, regulatory and policy infrastructure support is essential. The complete deregulation of the telecommunications sector, for example, can be instrumental in enhancing uptake rates, as costs are bound to come down enough for even the poor to gain access. The rapid uptake of mobile technology globally is a case in point. In 2007 alone, there were roughly 350m broadband Internet access accounts and 1.5bn mobile subscribers on the world's networks. Economist Intelligence EIU/ IBM Institute for Business Value (2008) estimated that the world would reach the 50% mobile penetration rate by mid 2008, and hit the 75% level less than four years' thereafter due to the deregulation of markets.

Policy and regulation is necessary to ensure that businesses and consumers are protected in their investments in the economy and for good overall economic performance. Governments are expected to allow market forces to build competitive telecommunications and Internet service markets. Governments' main role in a free market economy is to create an environment that would enhance equitable access to resources by network operators while ensuring that access to disadvantaged groups in a given population is also facilitated. A country's legal environment also provides the basis for free and fair commerce. This is especially critical when legitimising online transactions or admitting digital signatures in a court of law. Policy makers are supposed to create an environment in which digital connections can proliferate and where citizens and businesses find it convenient, efficient and profitable to use digital channels for their transactions.

Internet, WWW and Electrical Infrastructure

Over the past decade, developments in ICTs have transformed the world and turned it into a digitally networked community, with the Internet and the World Wide Web (WWW) gaining wider acceptance and use in government, academia and the corporate sector. There is great potential for the growth of the Internet and associated new technologies with regard to users, devices, speed and bandwidth, content, and applications, which are currently evolving and bound to have great implications in the information age.

The Internet has been covered extensively in many books; still, it is important to stress some of its qualities here, one being that it has provided businesses with a world without political and physical boundaries/barriers, making the global exchange of goods and services extremely easy. The Internet has been instrumental in the growth of the digital economy, especially SMEs, because it has increased transparency and global reach. Internet connectivity is also an important measure of e-readiness and an indicator of the extent to which individuals and businesses can partake in the digital economy. Increasingly, connectivity to the Internet is being defined by the extent to which personal computers, mobile-phone subscriptions, wireless Internet (WiFi) and broadband Internet accounts pervade a given user community. The penetration of secure Internet servers in a given population is also a good indicator of the extent to which reliable business transactions can be enabled or processed within and between businesses electronically. The impact on organisational businesses is manifested through intranets and extranets (internal and external communication networks, respectively), which are based on Internet Protocols (IPs) created to facilitate employees communicating with one another within organisations and between trading organisations. These Internet-based networks facilitate downloading and uploading information onto and from the company's computers. Extranets in particular are set up with major suppliers and distributors to facilitate information exchange, orders, transactions and payments.

The Internet has, since the late 90s, pervaded the public domain and altered in a significant way how businesses and consumers interact. As the Internet has evolved, so too has the digital economy. The evolution of the digital economy gained momentum from the year 2000 when online banking and finance, online shopping, online news, and e-mail communication - riding on new Internet developments - started to pervade the business sector, especially in the developed world. Today the Internet, which emerged as a rudimentary proprietary technology created by the US military in the 1960s, has grown in sophistication to become a global hive of interconnected computer networks, thus enabling many companies to take advantage of Internet business. It has been critical in globalising the world economy by creating free market capitalism, meaning that people with connectivity are able to buy and sell in cyberspace without restrictions. In short, the Net is a major component of the digital economy because it offers firms, individuals and governments an electronic network that enables the creation of virtual auction markets for goods and services where they previously did not exist. Secure servers on the Internet allow users to encrypt information, such as credit card data, thus facilitating e-commerce. Secure Internet servers also give an indication of the size and distribution of e-commerce (OECD, 2005).

The World Wide Web is part of the Internet, and consists of a system of servers that contain specially formatted information which can be accessed through the intervention of a web browser. It provides a way of accessing and sharing information on the Net by using the Hypertext Transfer Protocol (HTTP) to transmit data and allow applications to communicate or exchange business logic and share information. The WWW consists largely of websites, which can potentially enable SMEs to enhance their visibility to both current and potential clients. Through such websites, SMEs can also carry out Internet business transactions. Ramsey et al. (2003:256), in a study on the website use of SMEs in Ireland, observed that the most commonly cited reasons for establishing or maintaining websites were to: promote the organisation's name and intent; communicate specific product and/or service information; provide customer services; and advertise or market products/services.

The effective use of the Internet and WWW is predicated upon the sufficient and regular flow of electricity to power computers, networks and other technological infrastructure in businesses on a 24/7 basis. In some economies, especially in developing countries, electricity downtime is a major constraint on economic growth and development; without a cheap and reliable electrical power supply, most SMEs would fall behind and remain unconnected, making their operations in the digital economy increasingly impossible.

Hardware and Software Industries

Hardware is a key infrastructure in processing, storing and moving business content in the digital economy. Various types of hardware interoperate in the digital economy environment, some of which include microprocessor technology, parallel processors, neural networks, client/server technology, disk storage, open systems, user technologies, artificial intelligence systems, and virtual reality systems. Recent types of hardware proliferating in the digital economy are wireless and mobile communication technologies that enable consumers to place orders online through their cell phones and Personal Digital Assistants (PDAs).

The software industry is responsible for developing products and services that are consumed by SMEs in the digital economy. Such products and services include enterprise resource planning; customer relationship management systems; SQL servers; Oracle; SAP systems; computer programming services; pre-packaged software; computer integrated systems design; computer processing; data preparation; information retrieval services; and computer services management.

Telecommunications Industry

The telecommunications industry generates the communication products and services needed to transact business electronically. Such products and services include telephone to telephone communications; radio broadcasting; television broadcasting; and videoconferencing facilities/services. A further aspect of telecommunications infrastructure is Global Information Infrastructure (GII), the predecessor of the Internet. This is more than a mere 'network of networks' because it provides a worldwide assembly of systems that integrates a number of components, including communications networks such as the telephone, cellular phone and satellite networks; and information equipment/appliances such as computers, televisions, telephones, etc. GII is evolving into a mobile and ubiquitous computing system through which multiple devices - personal digital assistants, the Internet, home intranet, mobile phones, wireless laptops, cameras, printers, scanners, servers, etc - can integrate seamlessly using common interfaces and exchange information. GII also supports applications and services such as e-government services, e-commerce and telecommuting, to mention a few.

Digital Service Providers

A number of digital service providers are available in the digital economy environment, a few of which include Internet Service Providers, Internet Content Providers and Application Service Providers. Internet Content Providers (ICPs) exist largely as portal sites to provide content that is attractive to customers. Some commercial ICPs also facilitate online transactions (usually of the B2C type) to generate revenue. Other ICPs specialise in one single topic and provide one-stop information search services. These 'portal' sites came about naturally to manage the infinite information available on the Internet. For example, tdctrade.com is a leading trade portal created to facilitate Hong Kong companies' overseas trading activities. Application Service Providers (ASPs) are a specialised type of ICP that deliver and manage applications and computer services from remote data centres to multiple users via the Internet or a private network. They help registered companies outsource their IT functions so that they may concentrate on their core transactions. It is believed that SMEs can benefit most from ASP services. Other than technical support, ASPs can provide integrated e-commerce solutions, such as supply chain management, to clients.

E-Business and E-Commerce Industry

E-business and e-commerce are another critical component of the digital economy. The use of electronic platforms in a company's business transactions plays an important role in changing the way the company conducts its business. Through e-business, companies are able to engage in communication flows with local and overseas clients and agents more cost effectively. E-business also improves processing and customer response time through e-mail, EDI, the intranet and other such facilities. E-commerce - the process of buying and selling goods and services electronically with computerized business transactions using the Internet, networks, and other digital technologies (Laudon and Laudon, 2005) - also offers unprecedented opportunities, especially for SMEs in developing countries, for economic growth and development. E-commerce transactions may include placing orders, invoicing, the shipment of documents, marketing, market research, customer service, and finding potential customers and suppliers, to name a few. When compared to e-business, e-commerce is more specialised.

There are various types of e-business and e-commerce interactions. B2C (business to consumer) is the e-interaction between an enterprise and a customer; B2B (business to business) refers to e-business transactions between two companies; G2C (government to consumer) occurs when e-business transactions occur between government and consumers; and G2B (government to business) is when transactions occur between government and businesses. E-business and e-commerce are therefore facilitated by widespread access to computers and the Internet.

Information and Knowledge Management Systems

SMEs' ability to realize their goals depends on how well they acquire, interpret, synthesise, evaluate and understand information, and how well their information channels support organisational processes. Many SMEs may be interested in moving into e-business, but need additional information on key issues to take the next step (The Canadian e-Business Initiative, 2003). Lack of information and education can be the greatest barrier to SMEs wishing to partake in the new economy.

The digital economy is necessarily a knowledge economy, and is based on the application of human know-how. Knowledge management tools and human resources are necessary for SMEs to navigate the complexity of the digital economy. As both intermediate and consumption goods, information and knowledge add value and produce a return on investment; hence they must be harnessed and well managed. Corporate organisations are becoming more knowledge-intensive and operate in an environment where, for reasons of business competitiveness, knowledge sharing is widely valued. In a knowledge-intensive economy, goods and services are also increasingly produced in the form of intangible capital, making knowledge an important determinant in competitions between public bodies.

Intellectual Property Rights

Intellectual Property Rights (IPRs) are an important aspect of digital economy infrastructure. Balanced regulation in the use of IPRs is important for small-sized enterprises, especially with regard to the protection of ICT innovations and digital content products, and in mitigating the cost of obtaining, maintaining and enforcing intellectual property. Regulatory frameworks that balance the interests of suppliers and users are necessary to protect and manage intellectual property and digital rights without disadvantaging innovative e-business and content distribution models (OECD, 2004). Important aspects of intellectual property in the digital economy include understanding IP aspects of e-commerce; taking stock of a business's IP assets as they relate to e-commerce; taking cognisance of IP issues when designing and building websites; considering IP issues related to Internet domain names; analysing the effect of patents on e-commerce businesses; considering IP issues in the distribution of content on the Internet; IP contractual considerations; IP considerations in partnerships with government and educational institutions; and IP concerns about international transactions in e-commerce.

Human Capital and Knowledge Workers

Human resources are a crucial factor in the competitiveness of enterprises. Enterprises that lack skills are restricted in their application of new technologies and in their growth potential (Bojic, 2001). Human capital, the collective knowledge and skills of a firm's workforce, is therefore an important component in the digital economy as it largely determines the success or failure of businesses in a competitive environment. SMEs require a workforce that has ICT skills to enable them to compete in international markets. Human capital that includes IT specialists, especially with web-based expertise, is critical in determining an organization's e-readiness. In order to effectively participate in the digital economy, SMEs need to develop a strong human capital base. Human capital is reflected in the knowledge workers - described as employees - who manipulate symbols rather than machines, i.e. architects and bank workers, fashion designers and pharmaceutical researchers, teachers, policy analysts, etc (Varian, 1996). Knowledge workers are what Machlup (1962) described as individuals who produce information goods and services in industrial firms, institutions, departments and households, among others. The knowledge workers' levels of education, research and development, communication, and access to information services and information machines, encapsulate the scope of the information economy.

Research and Development (R&D)

This component of the digital economy is derived from the level of networking between various business entities, such as researchers (in universities, research institutions, etc); the private sector; investors in digital SMEs; business associations; public authorities; business consultants; business support agencies; customers; suppliers; government regulators; financial institutions; managers; employees; and the public at large. Digital communication and knowledge exchange naturally occurs between these entities, thus promoting projects for R&D in ICT enabled business services. R&D also involves launching IT incubators as a support service, mostly for new technology-based start-ups, which are an important source for digital economic growth.

Emerging Technologies

Emerging technologies are set to play an important role in the evolving digital economy. Recent emerging technologies that are gaining widespread use include Wi-fi, Wi-Max, Bluetooth and Voice Over IP (VOIP). Wi-Max is an abbreviation for

World Wide Interoperability for Microwave Access, and refers to broadband wireless access that is capable of transmitting network signals covering in excess of 30 miles in a linear service area. It provides shared data rates of up to 70 Mbps, which is also greater than Wi-Fi's theoretical equivalent of 54 Mbps. This technology has been useful in extending landline telephone connectivity. It is also applied in mobile devices, laptops, and other such hardware. Bluetooth technology allows intelligent, fast and cheap wireless communication between devices. Bluetooth devices do not require connection to any particular device, and recognise each other automatically. Moreover, Bluetooth facilitates roaming and can support wireless mice and web cams. VOIP, on the other hand, is Internet telephony that uses efficient and cheaper Internet protocols to route voice traffic over company networks. VOIP emerged because of the convergence of data networks and telephone services. It is possible to route several calls over a VOIP network, which is a vast improvement on the circuit switch used for traditional voice calls (Mutula and Wamukoya, 2007).

Working closely with emerging technologies are new applications based on the World Wide Web, known as Web 2.0. O'Really (2005) refers to Web 2.0 as "the network platform, spanning all connected devices and applications that make the most of the intrinsic advantages of that platform, delivering software as a continually-updated service that gets better the more people use it; consuming and remixing data from multiple sources, including individual users, while providing their own data and services in a form that allows remixing by others, creating network effects through an "architecture of participation," and going beyond Web 1.0 to deliver rich user experiences". Examples of Web 2.0 applications include digital commons, blogs, social networking sites, and wikis, to name a few. Other than this, the use of open source software is gaining momentum, especially in terms of providing platforms aimed at developing custom made applications for SMEs in the digital economy. Major benefits of open software include reduced costs and less dependency on imported technology and skills; affordable software for individuals; low cost licensing; and the ability to customize the software and use local languages.

CONCLUSION

This chapter has discussed the various components and infrastructures that make up the digital economy. Essentially, the digital economy environment is composed of government; policy and regulatory frameworks; infrastructure as it relates to the Internet, World Wide Web and electricity; the hardware and software industries; the telecommunications industry; digital service providers; e-business and e-commerce; information and knowledge management systems; intellectual property rights; telecommunications service providers; human capital and knowledge workers; research and development; and emerging technologies.

REFERENCES

Bojic, O. (2001). *Mapping SMEs policies in the digital economy-Germany, Greece, and the UK: A guide to readers*. London: Information Society Technologies.

Canadian E-Business Initiative. (2003). *Innovation in Canada*. Retrieved on November 16, 2003, from, http://www.cebi.ca/

Economist Intelligent Unit / The IBM Institute for Business Value. (2008). *E-Readiness rankings 2008*. London: EIU. Retrieved on October 25, 2008, from http://a330.g.akamai.net/7/330/25828/20080331202303/graphics.eiu.com/upload/ibm_ereadiness_2008.pdf

Laudon, K. C., & Laudon, J. P. (2005). *Management information systems: Managing the digital firm*. New Jersey: Pearson Education, Inc.

Machlup, F. (1962). *The production and distribution of knowledge in the United States*. Princeton: Princeton University Press.

Mutula, S. M., & Wamukoya, J. (2007). *Web information management: A cross-disciplinary textbook*. London: Chandos Publishing.

O'Really, T. (2005). *Web 2.0: Compact definition*. Retrieved on September 23, 2007, from http://radar.oreilly.com/archives/2005/10/web_20_compact_definition.html

OECD. (2004). *ICT, e-business, and SMES*. Paris: Publications Service.

OECD. (2005). *ICT key indicators, 20055*. Retrieved on August 7, 2008, from www.oecd.org/sti/ICTindicators

Ramsey, E., P. Ibbotson, J. Bell, and B. Gray (2003). E-Opportunities of service sector SMEs: An Irish cross-border study. Journal of Small Business and Enterprise Development, 10(3), 250–264. doi:10.1108/14626000310489709

Varian, H. R. (1996). The information economy: How much will two bits be worth in the digital marketplace? *Educom Review*, 44-46. Retrieved on March 18, 2007, from http://www.educom.edu

Chapter 4
SMEs Industry

INTRODUCTION

There is a growing impetus internationally for the development of strong SME sectors as engines of economic growth and development. This impetus is driven by the recognition of SMEs as the backbone of most economies. For instance, there are about 30 million enterprises in Europe providing employment to around 122 million people, with the vast majority (99.8%) of these enterprises being SMEs (Riegebauer, 2004). These SMEs therefore generate a large share of the GDP (Gross Domestic Product) in most of the region's countries. Asian countries are also investing significantly in SMEs. For example, during the 1980s and 1990s, China's economy was largely driven by SMEs, while the Philippines invested heavily in the same (Gungen, 2003). Moreover, the thriving economies of Asian countries such as Taiwan,

DOI: 10.4018/978-1-60566-420-0.ch004

Table 1. Contribution of SMEs to national economies (Source: UNCTAD, 1998)

Country	Share in the economy
Japan	99% of all establishments 52% of output 72% of employment 13% of exports
Taiwan	97% of all establishments 81% of output 79% of employment 48% of exports
Singapore	97% of all establishments 32% of output 58% of employment 16% of exports
Republic of Korea	90% of all establishments 33% of output 51% of employment 40% of exports
Malaysia	92% of all establishments 13% of output 17% of employment 15% of exports
India	95% of all industrial establishments 40% of industrial output 45% of industrial employment 35% of exports

Singapore, Thailand, and Malaysia have, for a long time, largely been driven by SMEs (Southern African Entrepreneurship and Small Business Association, 2002). In Malaysia, for example, SMEs are recognized as central to the country's economy and form a significant portion of the manufacturing and services sector. In December 2003, 89.8% of the more than 20,000 companies recorded in the manufacturing sector were SMEs, and the overall number of SMEs comprised 98.8% of the close to 193,000 enterprises in the services sector (Ramayah et al., 2003).

The economic potential of SMEs was reaffirmed at the World Summit on the Information Society in Geneva in 2003. The Summit's Declaration of Principles and the Plan of Action recognised SMEs as a special target group for broad-based economic growth (WSIS, 2003). SMEs are also vital for employment and a breeding ground for entrepreneurship. The diversity and impact of SMEs in any nation's economy is appreciated when its proportion in relation to a country's economy, employment capacity, exports, etc; are examined. Table 1 below presents the status of the SME industry in leading economies in Asia. Other regions of the world, such as Europe, North America and Africa, are discussed in subsequent sections.

STATUS OF SMEs IN EUROPE AND AMERICA

Internationally, small-sized enterprises are increasingly becoming the most vital part of nations' economies and are perceived to play a key role in fostering economic growth and creating jobs, thus alleviating poverty. SMEs account for 60 to 70% of the job market in most developed and developing countries (OECD, 2000). SMEs are also globally responsible for being the overwhelming generators of employment in all economies. In OECD member states, SMEs are fundamental for the economic growth in member countries and constitute for over 95% of all enterprises (OECD, 2000). For example, in the United States, United Kingdom, France, Japan and Germany, small-sized enterprises account for more than 60% of mainstream economic activity (Ntsala, 2000:171).

Within the larger European Union, SMEs are by far the most relevant actors in the economy because of their high contribution to employment and value added generation. They increase employment, thus compensating for the reduction in the labour-force undertaken by large enterprises (The BEEP Knowledge System, 2003). For instance, Western Europe's, SME industry accounted for 99.8% of all enterprises and 66.2% of all employment in 2004 (Christianson, 2004:45).

In Canada, the SME sector was responsible for a significant proportion of the jobs created and the value that was added to the economy in the 90's. By 2003, there were about 2.2 million SMEs in Canada, employing half of the private sector labour force (Canadian E-business Initiative, 2003). Like its Asian and European counterparts, the Canadian government also considers SMEs to be the backbone of the country's economy. Firms in this category represent 99% of businesses and account for over 80% of national employment and 85% of new job generation. In the United States, the average company that supports and promotes the economy belongs to the SME sector. 78% of the country's wealth is reportedly held by family businesses, and 12 million family businesses provide 59% of the country's total employment (Shevel, 2004:7).

SME SECTOR IN AFRICA

In keeping with developed and transitional economies, there is a growing impetus for the creation of strong SME sectors in most parts of Africa as engines for economic growth, supported largely by international lending agencies. SMEs in Africa employ more than 40% of all new entrants to the labour force, primarily because they tend to be more labour intensive than large firms and thus better able to mop up unemployment (Christianson, 2004:47). Countries such as South Africa, Egypt, Morocco, Kenya, Uganda, Botswana and Tanzania have prioritised their investment in SMEs (Gordon, 2003:15; Maksoud and Youseff, 2003:1-15 and Republic of Botswana, 2003). Africa's informal sector is huge because of the large numbers of employees it has absorbed as a consequence of inadequate jobs and opportunities in the formal sector. The informal sector in many African countries is estimated to employ from three to six times the number of employees in the formal sector (Muuka, 2002). SMEs in East African countries such as Kenya, Uganda and Tanzania, provide employment to more than 50% of the labour force and account for more than 50% of the GDP in the region (Chowdhury and Wolf, 2003). In Kenya alone, SMEs contribute about 12-14% to the country's national income.

The Egyptian government ranks SMEs top amongst its priorities of socio-economic development. The SME sector in Egypt constitutes more than 99% of all non-agricultural private enterprises and generates 75% in terms of employment (Maksoud and Youseff, 2003). The Southern African Development Community (SADC) region, consisting of Angola, Botswana, Democratic Republic of Congo, Lesotho, Malawi, Mauritius, Mozambique, Namibia, Seychelles, South Africa, Swaziland, Tanzania, Zambia, and Zimbabwe; also places a high premium on the SME sector as an engine for economic growth in the member states. This high regard for SMEs has been borne out of the realisation that the world economy has undergone phenomenal changes over the last few decades, occasioning continued advancement in the industrialisation process and an increase in nations' wealth (SADC, 1999:1-64). SADC member countries also appreciate that due to the growing globalisation of the world economy, there has been internationalisation with regard to the production of goods and services, leading to the growth of massive multinationals that dominate world trade. SADC member states therefore recognise that they must take appropriate action, including investment in the SME sector, to take part in global business markets.

The South African President, in his state of the nation address in May 2004, signalled his government's commitment to the promotion of a robust SME sector. The President noted that major barriers to small-sized enterprises development could be removed through greater policy coherence and implementation measures within and between the various relevant government departments and sectors of the economy. In particular, the President stated that the regulatory framework for SMEs would be reviewed by the end of September in 2004 to provide an enabling environment for a thriving SME sector. Other changes to the sector that the President promised would be implemented included merging Ntsika - the agency for assisting and empowering SMEs - with national manufacturing advisory centres to create a unified development agency. In addition, he stated that the government would set up a micro credit agency to enable easy access to start-up capital.

The South African small-sized enterprises sector is considered to be the largest employer in the country, with approximately 60% of all employment in the private sector being provided by SMEs (Ray, 2004:13). The South African government estimates that the SME sector will contribute 35% to the country's GDP by around 2013 - 2018 (Gordon, 2003). The Anglo American mining company (a major South African conglomerate), through its empowerment and support initiatives, spent more than 11 billion Rand (US$ 1.1 bn) in 2002 on Black owned businesses and Black business development. Through this initiative, the mining company supported businesses by helping them borrow money from banking institutions, and also provided equipment, vehicles, management, training support and the transfer of business skills. The company further facilitated networking between SMEs and larger companies and provided access to direct foreign investment by SMEs. Entrepreneurs also benefited from access to Anglo's infrastructure. For instance, the legal department could be called upon in the unlikely event that a legal problem arose within an SME. South Africa therefore exemplifies a perfect environment for a thriving SME sector in the SADC region in particular and the whole of Africa in general. It is estimated that 80% of the SMEs in the country are family owned businesses, and that South African SMEs employ about 54% of the country's total workforce in the private sector (Shevel, 2004:5).

In Botswana, the government also ranks the SME sector high amongst its development priorities. SMEs are perceived as critical in creating employment and increasing the country's competitiveness both internationally and nationally. The government also recognises that as a result of globalisation, even small-sized enterprises should find it increasingly easy to source and deliver products and services to any place in the world, and thus gain access to new markets and new opportunities (Republic of Botswana, 2003). Botswana, like many other developing countries, faces the challenges of unemployment and poverty, and for this reason, the country's government and donor agencies have increasingly come to emphasize the central role that the SME sector can play in promoting economic and social development. A lot of financial assistance has consequently been channelled to encourage and empower SMEs in the country (Machacha, 2002:277). The government has also realised that it is not sensible to depend on mining for the long-term economic development of the country. Consequently, it is encouraging the development of the SME sector as a way to diversify its economy. For instance, in 2001, the government established the Citizen Entrepreneurial Development Agency (CEDA), a micro-credit financial entity that facilitates access to finance by citizen-owned enterprises, particularly SMEs.

Efforts aimed at promoting the SME sector in Africa are being facilitated through various organisations. The Commonwealth Business Council, in conjunction with the International Telecommunication Union, the Government of Mauritius, and the e-Africa Commission of NEPAD (New partnership for Africa's Development), have in the past taken steps to promote SME development on the African continent. In July 2004, they jointly organised an ICT stakeholders' Forum in Johannesburg, South Africa, which focused on less developed countries and examined concrete projects, proposals and models that would help integrate less developed countries into the global economy through the effective deployment of ICTs and SME development (Banerji, 2004). The World Bank also supports and encourages strong SME sectors in Africa. The bank has recognized that in order for Africa to achieve sustainable economic development, efforts must focus on the entrepreneurial base in each country

(Bindra, 2004). Similarly, in July 2004, the United Nations Development Programme (UNDP) launched the Africa-Asia SME Network Programme in Johannesburg, known as TECHNONET Africa. At that time, TECHNONET consisted of 12 SME public and private sector institutions from 7 African countries, namely Cameroon, Ghana, Mozambique, Nigeria, South Africa, Tanzania and Uganda. It stemmed from the Tokyo International Conferences for African Development (TICAD) that took place in 1993 and 1998 respectively. The establishment of TECHNONET was based on the critical role played by SMEs in the creation of employment opportunities and income generation to reduce poverty in Africa (Senior Human Resources Associate, 2004:44).

BARRIERS TO THE SMEs INDUSTRY: CASE STUDY OF THE AGRICULTURAL SECTOR IN AFRICA

Within Africa, the agricultural sector remains the mainstay of most nations' economies, with small-sized enterprises farmers constituting the majority of the sector. Most countries also rely on agricultural products for exports and for meeting their own food needs. However, SMEs face challenges related to information access; the digital divide; poor mechanisms for technology transfer; restrictive intellectual property rights (IPRs); and lack of access to micro-credit. Most crops exported from Africa are less competitive than those from Europe and North America, in part due to the subsidies enjoyed by farmers in developed nations, which African farmers do not necessarily get from their own governments.

Moreover, small-sized enterprises farmers in the developing world often find their crops affected by diseases and low or unstable pricing. This situation is exacerbated by the fact that the operation of commodity markets tends to favour large scale buyers who have better access to technology and information, and who are also closer to the markets (Dench, 2001). African small-scale farmers are also adversely affected by global imperatives; as Felsenstein (2005) observes, global "free trade" policies open local markets to foreign-produced agricultural products and destroy the basic source of income of farming families. Furthermore, these policies seem to work only one way – agricultural industries in the developed world strenuously resist opening their markets on the grounds that prices would drop. Thus, the effects of this policy of freedom (for those who have the necessary wealth and political might) are forced upon those who only have access to their own land and labour (Anyanwu, 1996).

To operate fairly, the ideal free market requires free access to information. However, many SMEs in the agricultural sector in Africa face significant challenges relating to information access. For people without access to telephones or electricity, it is impossible to make informed decisions about which market in town they should take their crops based on the competitive prices on offer. Without the ability to obtain this kind of competitive information, farmers, as Felsenstein (2005) observes, do not become active participants in markets, but rather resources that are mined and eventually discarded. The type of information that is available to SMEs is likely to influence the decision by buyers and sellers to engage in business to business (B2B) trading hubs. Rizk (2004), in a study on Egyptian SMEs in the textile industry to assess the firms' levels of connectivity, awareness and use of ICTs in marketing, production, and management, concluded that ICTs offered many potential benefits for SMEs, listing some as: reduced transaction costs; opportunities to overcome the competitive advantage of larger firms (as determined by economies of scale); opportunities for innovation; and the emergence of new products and services.

The current digital economy calls for information to be at the centre of all decision making activities. However, Africa faces obstacles in ICT development, especially with regard to slow reforms in the telecommunications sector as characterized by poor infrastructure, scarce human resources, and the poor quality of services and paucity of information. Small-sized enterprises are at a greater disadvantage because of their inability to pay for the high costs of communication, their lack of access to technology, and inadequate support from government. In Africa, 70% of Internet traffic is routed outside the continent, significantly driving up the costs of (Internet) access for businesses and consumers. This means that a great number of businesses are unable to either get or sustain connectivity. On top of this is the shortage in the power supply that has hit most parts of Africa, especially countries in Southern Africa, East Africa and parts of West Africa. In South Africa, for example, extreme demands on the electrical system have at times resulted in power interruptions or what is popularly known in the region as load-shedding. Current indications point to a tight balancing act between supply and demand that will cause problems in the near future (Botswana Power Corporation, 2008).

TECHNOLOGY TRANSFER CHALLENGES OF SMEs IN THE AGRICULTURAL SECTOR IN AFRICA

While some of the challenges facing SMEs in the agricultural sector are of national concern, others are of global dimensions. In the evolving digital economy, access to technology is the foremost determinant of success. However, SMEs often fail to adopt technology for various reasons, including lack of applicability and little incentive to change business models when returns are unclear.

SMEs also face generic barriers to the adoption of technology, such as trust, transaction security and IPR concerns, and challenges in areas of management, technological capabilities, productivity, and competitiveness (OECD, 2004). They also tend to be concerned with the legal uncertainties of adopting technology, especially in relation to the application of e-commerce in contracts, terms of delivery and guarantees, cross-border transactions or distance contracts, and the validity of electronic signatures. Whereas larger enterprises can call upon their (in-house) lawyers, SMEs do not have the necessary expertise or the financial resources to master complex and varying legal rules. Moreover, security aspects of the Internet create distrust, as do national differences in consumer protection legislation. Uncertainty over payments, especially with respect to concerns about the security and confidentiality of online payments, also often impede on the adoption of e-commerce (Riegebauer, 2004).

In Africa, the agricultural sector is largely dominated by small-sized enterprises farmers who are involved in subsistence farming with limited access to appropriate resources because they are constrained by poverty and receive little assistance from government (Koch, 2004). The lack of effective land policies constrains SMEs in terms of access to technology, land, capital, expertise, and good marketing infrastructure. Consequently, despite the fact that Africa is endowed with huge arable land, water and animal resources – enough in fact to make it the world's leading food producer - agriculture is characterised by low productivity and low returns. Effective land policies would, for example, include: security with regard to land ownership; capacity building for the effective use of the land; access to credit in order to buy land; access to farm technologies; provision of agricultural subsidies; and support services for land use.

The African agricultural sector suffers from poor or inadequate technologies - tractors, combine harvesters, fertilizers, insecticides, pesticides, weed killers, etc - due to a lack of finance. The existing technologies are also underutilized for a number of reasons, such as lack of access to spare-parts, the lack of skills necessary to repair things when they break down, and illiteracy. Moreover farmers, especially those in the small-sized enterprises farming industry, often lack the capacity to use farming technologies. This is often because they do not have access to information about various farming technologies, how they work, and how they can benefit from them. African small-sized farmers are also unable to acquire modern technology because of reasons that Woherem (1993) lists as: stringent proprietary IPR conditions attached to the use of technologies; implantation rather than technology transfer from the developed world; lack of enabling policies; and biased contractual arrangements in favour of technology companies that are based in the developed world; the lack of policies that effectively promote the transfer of technological know-how to the farmers; difficulty in acquiring information related to science and technology; lack of capital intensive technologies, making farmers largely reliant on labour intensive technologies that limit mass production; and the lack of understanding on how best to procure IT.

As mentioned, African countries generally lack enabling legislative and regulatory mechanisms to promote the transfer of technology from the developed or other developing worlds that make them. Little effort has been directed towards accessing private sector technological resources within individual countries (African Technology Foundation, 2002). Furthermore, African research programmes have often been developed in isolation; as a result, a technology developed in one African country may not get used elsewhere because the infrastructure necessary to transfer it is non-existent. Toure (1996), in a status report on agricultural technology development, transfer and commercialization in six African countries, found that although progress had been made, the continent remained significantly behind the developed world. He noted that for Africa to increase its agricultural productivity in order to attain food security, reduce poverty, provide employment, increase its exports and produce raw materials for industrial transformation, the commercialization of modern agricultural technologies is imperative.

Africa also suffers from a paucity of scientific research to facilitate effective technology transfer. For example, the safety of Genetically Modified (GM) technologies in the transfer of genes into local planting material has not been effectively researched (Koch, 2004). Consequently, it is difficult to develop bio-safety measures that help identify risks with regard to GM crop management. This is exacerbated by small-sized enterprises farmers' suspicion that multinational companies involved in biotechnology, such as Syngenta and Monsanto, are making a killing in terms of their return on investments linked to intellectual property regimes. As a result, countries in the developing world are wary of the proponents of, or the multinational companies involved in, the transfer of GM technology.

Africa also suffers from a paucity of bio-safety laws and review systems able to assess the safety of GM trials, provide permission for such trials, and monitor their compliance with necessary risk management systems (Morris and Koch, 2002). Because of the absence of effective bio-safety laws, biotechnology is mired in the controversy that surrounds its viability and environmental impact. The lack of effective legal systems may also lead to the abuse, misuse or inappropriate application of GM products, as would poor understanding of biotechnology. There is also a general lack of understanding of IPRs associated with GM technology development by many African governments (Koch, 2004).

INTELLECTUAL PROPERTY CHALLENGES OF SMEs IN AFRICA

African farmers' difficulties in terms of access to technologies are compounded by the complexities associated with the protection of intellectual property rights. Patents and plant breeders' rights on protected materials often raise the cost of accessing new plant varieties, and this especially affects smallholder farmers in Africa who have almost no bargaining voice to rise against the multinational patents and IPRs holders of the western world (Omanya et al., 2005). The restrictive nature of IPRs' protection limits the saving, exchange and selling of farmer-produced seed of the protected variety, while access to such protected resources and innovated inputs through license and contract fees creates significant financial challenges for subsistence farmers. Farmers in Africa are used to saving and storing seed for the next season, and also share seeds with neighbours. However, GM products do not allow saving or sharing. Moreover, because of the monopoly of IPR associated with GM products, they are very expensive to small-sized enterprises farmers, especially with regard to the costs of chemicals and seeds.

Before the 1980s, IPR was principally an instrument of national policy, meaning that countries tailored their IPR regimes in accordance with national economic and social conditions. But now, IPR has become a policy issue subject to international practice, where countries that fail to conform to new global standards are threatened with retaliation in the form of losing trade privileges (Shadlen, 2007). The inclusion of Trade Related Intellectual Property Rights (TRIPs) in the WTO international trade policy encourages poorer countries to adopt IPR systems that are inappropriate for their level of scientific and technological development. This is because TRIPs protects richer countries' markets while encouraging the liberalization of poor countries' markets. The legal instruments governing the global IPR regime were designed with a focus on the commercial sector, and do not take into account the situation of resource-starved farmers who produce, save, exchange, replant and sell planting resources at local level (Wekundah, 2004).

For Africa to make headway in dealing with IPR challenges, the effective and responsible management of IPRs should begin with the formulation of appropriate policies in African member states that: set clear objectives and principles in obtaining access to, and the use of, such IPR and protected technologies; establish guidelines on how and when IPR protection can be sought and exercised; promote basic principles concerning the use of IPR and protected material by recipients; ensure rigorous and sustained negotiations with the developed world to ease trade barriers that inhibit African countries from effectively exporting to Europe and the US in particular.

MICRO CREDIT CHALLENGES OF SMEs IN AFRICA

Small-sized enterprises in Africa face many challenges in their attempt to obtain credit from financial institutions on account of lack of business skills and experience, lack of collateral, and the poor quality of business plans, among other factors. The importance of micro-credit to assist small-sized enterprises is widely acknowledged. Kofi Anan, the former UN Secretary General, stated in a message at the Micro credit Summit at the Africa Advocacy Forum II (held at the UN headquarters in New York between the 10-13th of November, 2002) that micro-credit is a critical anti-poverty tool and a wise investment in human capital. He pointed out that now that the nations of the world have committed themselves to halving, by the year 2015, the number of people living on less than a dollar a day, it was imperative to look even more seriously at the pivotal role that sustainable microfinance could play, and is playing, in reaching this Millennium Development Goal (Anan, 2002). It was recommended at the summit that a unified front of African microfinance institutions, researchers, trainers and financial supporters is needed to help African practitioners develop viable financing models for SMEs.

CONCLUSION

The agricultural sector in Africa provides a perfect example of the challenges that face SMEs and the impact that the digital economy can bring to bear on their operations. SMEs face various challenges relating to information access, technology transfer (conventional and GM), intellectual property rights and access to micro-credit that hinder their participation in the global market. These challenges, in turn, have policy and managerial implications. In terms of information access, policies that enhance the liberalization of the telecommunications sector to encourage competition and the modernisation of infrastructure (especially broadband networks) are necessary implementations. Universal access interventions are also necessary to facilitate access to ICTs and electricity. Africa could benefit if efforts were made to transfer technologies from private/public sectors, and if enabling policies and regulatory frameworks were put in place for the transfer of technology from within and between countries. Moreover, incentives that allow the private sector to lead in innovative technological activities are required. Efforts could also be made in the area of IPRs, especially with respect to their effective and responsible management. Governments have to provide effective policy and regulatory mechanisms that enable SMEs to access micro-credit from banking institutions without demand for collateral. Until these challenges are addressed, Africa is bound to remain less competitive in the emerging global digital economy.

Governments can provide a support plan to enable SMEs to reconstruct business processes, develop new businesses, and reduce costs of access to technology. Governments can also assist by providing training that helps SMEs cope with rapid advances in IT; improving SMEs technological power; exploring new sales channels; and helping SMEs add more value to their products and services. Managers of SMEs should prioritize the provision of information literacy to employees, especially with regard to the use of technology to access global markets. Finally, governments should work hand in hand with SMEs' managers to enhance their e-readiness. E-readiness efforts would focus on enhancing access to modern technological infrastructure, developing competent ICT skills, enabling access to credit, and showing SMEs how to transact business processes online.

REFERENCES

African Technology Foundation. (2002). *Business plan*. Retrieved on January 20, 2008, from http://www.aatf-africa.org/bplan.php

Anan, K. (2002). *Microcredit–a solution for Africa. Africa advocacy forum II*. New York: United Nations.

Anyanwu, L. O. (1996, April 22). Promoting equity markets in Africa. In *Proceedings of the Ninth International Conference Montclair State University.*

Banerji, S. (2004). *ICT stakeholders forum (Mauritius)*. Retrieved on April 22, 2004, from http://www.cbcglobelink.org/cbcglobelink/events/ICT04

The BEEP Knowledge System (2003). *The digital SME*. Brussels: EC.

Bindra, S. (2004). Let's give legitimacy to the small entrepreneur. *Sunday Nation*. Retrieved on January 25, 2004, from http://www.nationaudio.com/News/DailyNation/Today/Comment/NewsAnalysis2501200420.html

Botswana Power Corporation. (2008). *Managing the supply-demand balance*. Gaborone: Botswana Power Corporation.

Canadian E-Business Initiative (2003). *Innovation in Canada*. Retrieved on November 16, 2003, from http://www.cebi.ca/

Chowdhury, S. K., & Wolf, S. (2003). *Use of ICTs and economic performance of SMEs in East Africa. Discussion paper no. 2003/06*. Retrieved on November 29, 2003, from http://www.ciaonet.org/wps/chs08/chs08.pdf.

Christianson, D. (2004). ICSB-much more than just a talk shop. *Enterprise*, 47-48.

Dench, D. J. (2001). *Spilling the beans*. Retrieved on January 17, 2008, from http://www.fairtrade.org.uk/products.html

Felsenstein, L. (2005, March 15). *Fair-trade technology-tech that helps the world*. Paper presented at the O'Really Emerging Technology Conference, San Diego, CA.

Gordon, G. (2003, February 16). SME survey. *Sunday Times Business Times, 15*.

Gungen, T. (2003). *E-finance for SMEs in Turkey: Yapi Kredi Bank*. Retrieved on November 27, 2003, from http://r0.unctad.org/ecommerce/event_docs/xmefinoct2001/gungen.pdf

Koch, M. (2004). *The case for GMOs in the developing world-how African farmers are benefiting from biotechnology*. Retrieved on January 26, 2008, from http://www.cropscience.org.au/icsc2004/symposia/3/8/1834_koch.htm

Machacha, L. (2002, April 3-6). Impact of information technology on small and medium enterprises (SMEs) in Botswana. In *Proceedings of International Conference*, Port Elizabeth, South Africa (pp. 277-282).

Maksoud, S. S., & Youseff, M. A. A. (2003). *Information and communication technology for small and medium enterprises in Egypt*. Cairo: SME Development Unit-Ministry of Foreign Trade.

Morris, E. J., and M. Koch (2002). Review: Bio-safety of genetically modified crops-an African perspective. AgBiotechNet, 4, 102.

Muuka, G. N. (2002, April 3-6). Africa's informal sector matters: A challenge to scholars to close the knowledge gap. In *Proceedings of International Conference*, Port Elizabeth, South Africa (pp. 1-6).

Ntsala, M. (2000, October 26-27). Information delivery systems for small business. In *Proceedings of First Biannual DISSAnet Conference*, Pretoria (pp. 167-176). Pretoria: The Centre for Information Development, University of Pretoria.

OECD. (2000, June 14-15). Realising the potential of electronic commerce for SMEs in the global economy. *Conference for Ministers Responsible for SMEs and Industry Ministers*, Bologna, Italy. Retrieved on August 8, 2004, from http://www.conferenzabologna.ipi.it/Eng/dopo_la_conferenza/documenti/WK3_eng.pdf

OECD. (2004). *ICT, e-business, and SMES*. Paris: Publications Service.

Omanya, G. R., Boadi, F., Nang'ayo, M. H., & Bokanga, M. (2005, February 21-24). *Intellectual property rights and public/private partnerships for agricultural technology development and dissemination*. Paper presented at the Kenya National Conference on Revitalizing the Agricultural Sector, Nairobi, Kenya.

Ramayah, T., M. Jantan, R. Mohd Roslin, and R. Siron (2003). Technology readiness of owners/managers of SME's. International Journal of Knowledge Culture Change Management, 3, 475–486.

Ray, M. (2004). The SME agenda. *Enterprise, 13*.

Republic of Botswana. (2003). *Budget speech 2003*. Gaborone: Government Printer.

Riegebauer, R. L. (2004). *Self-regulation of e-commerce in Europe–time to think small: Building SMEs trust and confidence in the electronic marketplace*. Retrieved on November 16, 2008, from http://www.uu.nl/uupublish/content/Timetothinksmall.pdf

Rizk, N. (2004). *E-readiness assessment of small and medium enterprises in Egypt: A micro study*. Cairo: American University. Retrieved on August 9, 2004, from http://www.sba.luc.edu/orgs/meea/volume6/Rizk.htm

SADC. (1999). *SADC in the next millennium: The opportunities and challenges of information technology*. Gaborone: SADC Secretariat.

Senior Human Resources Associate. (2004, July 23-29). Africa-Asia small medium enterprise (SME) network programme. *Mail and Guardian, 44*.

Shadlen, K. C. (2007). Intellectual property, trade, and development: Can foes be friends? Global Governance, 13(20), 171–177.

Shevel, A. (2004). Family-run companies lead the corporate pack. *Sunday Times Business Times*, 4-5.

. (2002). The promotion of small business in SADC. Traders Journal for the Southern African Region, 8, 28–29.

Toure, M. (1996, September 4-6). Commercialization and transfer of agricultural technologies in sub-Saharan Africa. *A Roundtable Workshop University of Maryland Eastern Shore Princess Anne,* Maryland.

UNCTAD (1998). *Handbook on foreign direct investment by small and medium-sized enterprises: Lesson from Asia*. Geneva: UNCTAD.

Wekundah, J. (2004). *Intellectual property, partnerships, and people*. Nairobi: Biotechnology Trust Africa (BTA).

Woherem, E. E. (1993). *Information technology for Africa: Challenges and opportunities*. Nairobi: African Centre for Technology Studies (ACTS).

WSIS. (2003). *Draft declaration of principles-Geneva WSIS Civil Society Working Group*. Retrieved on November 7, 2004, from http://www.wsis-pct-org/prepcom2-declaration.html

Chapter 5
Digital Divide and SMEs

INTRODUCTION

There is a growing interest in how the digital divide affects the operations of SMEs. The digital divide among businesses is defined as the disparity between the effective use of ICT for gains in productivity (Wielicki, 2008). Often, SMEs are found to be on the wrong side of this divide, with insufficient funds and/or access to ICT as the most cited reasons for falling behind big corporations. A further common barrier to the use of ICT is cited as the lack of education and training. A study conducted on micro and small-sized enterprises (SMEs) from selected regions in Spain, Portugal and Poland on the use of ICT produced interesting results. It was found that the main barrier to the utilization of ICT and e-Business, and thus the main reason why SMEs struggle with the digital divide, is not so much the lack of access to IT as the

DOI: 10.4018/978-1-60566-420-0.ch005

lack of proper knowledge, education and skilled owner-managers and employees within the enterprise (Arendt, 2008).

Riccardini (2003), during the UNCTAD Expert Meeting on Measuring Electronic Commerce in Geneva in September, 2003, underscored various policy actions needed to bridge the digital divide among SMEs. He suggested that actions were particularly necessary in the following areas: policies that govern businesses, households and government in the use of ICTs; economic sector policies; infrastructure development, especially broadband; e-government; and e-content and territorial development policies. The extent of the digital divide among SMEs can be established using various indicators, such as ICT usage, level of e-commerce and e-business adoption, the e-public sector and e-government statistics. Such statistics would include e-commerce and e-business indicators such as purchases and sales of assets or services that occur through the Internet, Electronic Data Interchange (EDI), mobile phones or other computer-mediated networks Other indicators would include the goods and services ordered over those networks directly by an enterprise, although the payment and the delivery of the goods or services may be conducted on or offline.

Arbore and Ordaini (2006) investigated the digital divide factors affecting SMEs with regard to broadband access, and confirmed that the geographical situation of SMEs can be an obstacle, especially for small-sizedenterprises. In such instances, specific ICT organizational strategies - namely outsourcing solutions - are an important mitigating factor for size and location-based broadband disadvantages. Elsewhere, the Small and Medium Enterprise Administration Ministry of Economic Affairs in Taiwan identified some of the measures needed to bridge the digital divide among the SMEs in that economy as (Lin, 2007): e-commerce empowerment; recruiting an information society (IS) team; selecting e-products; delivering onsite services; creating IS revenue; providing e-commerce training; developing industry specific applications; developing e-applications; and promoting broadband.

A book launched in Costa Rica, written by Monge-Gonzalez, Alfaro-Azofeifa, and Alfaro-Chamberlain (2005) on the access, use, adoption and impact of ICTs on the productivity and competitiveness of micro-, small- and medium-sized enterprises (SMEs) in Central America, highlights how issues of the digital divide are obstacles to small-sized enterprises participation in the Net economy. In particular, the authors observed that access to ICTs is quite limited in the Central American Region, with a general lack of telephone lines, mobile phones and Internet connectivity. The authors therefore concluded that in order to make SMEs in Central America more productive and competitive in a connected economy, there must be a drive to increase access to ICTs and improve their adoption by SMEs through training and implementation programs.

THE DIGITAL DIVIDE PHENOMENON AND DEBATE

The digital divide remains the greatest obstacle for SMEs that wish to take their business transactions online. The debate about whether the digital divide between developing and developed countries is narrowing or widening has intensified over the last five years. Proponents of the view that the digital divide is narrowing between developed and developing countries cite Africa's leadership in mobile phone growth in the world. Those with a contradictory view often point out that access to technology is positively correlated to economic development and wealth creation. They argue that since the dawn of the last century, the gaps in wealth and poverty within and between developed and developing countries have continued to grow. More recently, they have come to highlight the growing divide between people in society with broadband access and those without, and also cite the disparity between those in society with access to fiber optics (faster broadband) and those on slower broadband networks (coaxial, copper wires, satellites). The digital divide, seen from the perspective of broadband access and quality, has acquired the nomenclature, 'new digital divide'.

Citizens Online (2007) is of the view that the digital divide is wider than ever, with those stuck on the wrong side getting more deeply excluded and harder-to-reach than ever before. The United Nations 2005 report (United Nations, 2005) is in agreement, observing that the spread of information technologies to a select group of the world's people was worsening the disparities between the 'e-haves' and the 'e-have-nots', thus fuelling the danger that this unequal diffusion of technology, far from its ideal of fermenting cohesion by providing opportunity, would result in reinforcing traditional patterns of economic and social inequalities that could lead to the weakening of social bonds and cultural organisation. During the 2003 World Summit on Information Society (WSIS) in Geneva, Kofi Anan, the former Secretary General of the United Nations, lamented that for too many people, the gains that can be made out of ICT remain out of reach. Anan also recognized a tremendous yearning not for technology per se, but for what technology can make possible (WSIS, 2003).

The International Telecommunication Union (2005), in its publication *ICT Opportunity Index* in time for the second WSIS meeting in Tunis in 2005, also showed quite clearly that digital opportunities were unequally distributed across developed and developing countries, and suggested that the gap between the ICT-poorest countries and most others is actually growing. The publication quite literally concluded that the 'have' and 'have-not' countries are worlds apart. The G7 group of countries, in a conference on *Information Society and Development* in South Africa in May 1996, concluded that a large gap existed between industrialized and less industrialized countries with respect to information infrastructure. At the time, the gap was occasioned by far lesser investment in ICT infrastructure by developing countries.

In a report entitled *African Region Communications Infrastructure Programme* released in April, 2007 (Nyasato and Kathuri, 2007), the World Bank observed that the East and Southern African regions suffer bandwidth deficiency because they account for less than 1% of the world's international bandwidth capacity. Additionally, countries in the regions lack direct terrestrial access to global information and communications infrastructure, relying on expensive satellite connectivity to link up to the rest of the world. As a result, users of telecommunications are faced with some of the highest costs in the world. To illustrate this, the report notes that the international wholesale bandwidth prices in the two regions are 20 to 40 times higher than in the United States, and international calls are on average 10 to 20 times more costly than in other developing countries. Moreover, Eastern and Southern Africa lack a fibre optic cable system. Consequently, the region relies exclusively on satellite links for voice and data transmission, at about ten times the cost and at transmission speeds of less than a quarter of what they'd enjoy with fibre optic links (Morris, 2007).

Despite this, the World Bank also believes that "the "digital divide" between rich and poor nations is narrowing fast, calling into question a costly United Nations campaign to bring hi-tech telecommunications to the developing world". According to the World Bank, because of the explosive growth of telecommunications services in poor countries, the digital divide is narrowing. The World Bank report was based on the premise that people in the developing world were getting more access, especially to cell phone communications, far faster than they got access to new technologies in the past. The report further noted that half of the world's population now enjoyed access to fixed-line telephones, and 77% to a mobile network - surpassing a WSIS campaign goal that called for 50% access by 2015 (Atkins, 2005).

The World Information Society Report for 2007, in advancing reasons for why the digital divide between developed and developing countries appears to be narrowing, stated that:

- Developing countries such as India and China are gaining on OECD countries in fixed line and mobile phone penetration
- Less Developed Countries (LDCs) are catching up with developed countries in mobile phones, Internet usage and broadband
- Internet usage is increasing. For example, in 1997 low income countries accounted for 5% of global Internet usage, and by 2005 this figure had risen to 30%

- Africa is currently the leader in mobile phone penetration
- There has generally been increased Internet growth in many countries worldwide.

Although the developing world, particularly Africa, has experienced tremendous progress in mobile phone penetration, the continent remains behind with regard to PC and fixed line penetration, quality of access, amount and relevance of content, affordability of access, and IT usage. As far as affordability is concerned, a study by the Business Leadership Group on 15 countries worldwide found that ADSL (broadband) costs in South Africa were 139% higher than the average rate in the nations surveyed. The study noted that local calls at peak hours were 199% more expensive (Naidoo, 2007). Consequently, if the cost of access in South Africa is taken as the base measure of the rest of the continent (given that South Africa has well developed telecommunications infrastructure), then the situation in other countries can only be worse. This is confirmed by the World Bank, as cited by Nyasato and Kathuri (2007), when it observed that international phone calls and high-speed Internet connection charges in 25 East and Southern African countries were beyond the reach of most people belonging to the middle or lower income groups. When compared to the developed world, sub-Saharan Africa also has a lot of catching up to do with regard to digital literacy, universal access and universal service.

The digital divide is exacerbated when there is an inadequate electricity supply, which is necessary to power up computer operations, as faced currently by a number of African countries. The March 2007 report of the NEPAD Support Unit of Economic Commission for Africa (ECA) noted that electricity use per capita on the African continent was less than 2%. The report concluded that without access to sufficient, good quality and reliable energy, every social and development activity is critically constrained. Evidently, in order to achieve the MDGs in a sustainable manner, Africa needs to address the unique challenges that energy deficits create. African countries must also pursue initiatives that would secure access to energy for at least 35% of Africa's population within 20 years, especially in rural areas (Economic Commission for Africa, 2006).

UNPACKING THE DIGITAL DIVIDE

The concept of the digital gap (divide) can be traced back to the information technology ideology that emerged in the United States in the 1970s, when conservative politics started to embrace liberal, laissez-faire free market values and technological determinism (Birdsall, 1997). The conservatives were motivated and influenced by the convergence of computing and telecommunications, which was perceived to be a technological foundation for economic growth. Later, the fusion of information technology with free market economics was adopted by major political parties in the United States. By the mid-90s, the Democratic Party captured and exploited the digital divide as a policy issue when Vice-President Albert Gore became a prominent advocate for the development of a National Information Infrastructure (NII) as an economic growth strategy. The concept of the digital gap (now divide) therefore evolved in the corridors of power of the political elite, and is consequently the creation of the highest echelons of society. Civilrights.org, (2005) quoting Wade Henderson, noted that "the digital divide debate is not about gadgets or even markets. It is a debate about who gets to speak and to hear, and for what price, and to whom".

Other than the political dimension of the digital divide, the phenomenon, as already pointed out, is not just about access or lack of access to ICTs. Green (2000), citing data from the *Education Week (2000)* teacher survey, observed that when teachers were asked why they do not use software or the Internet for instruction, they cited the following reasons, which have nothing to do with access:

- There is inadequate time to prepare and preview software or websites
- There is a lack of training on software
- It takes too much time to use technology
- The schools' computers are not powerful enough
- Technologies are not aligned with curricula and assessment
- It is difficult to find software that meets students' needs.

There appears to be a perception-reality gap between those who believe that the digital divide is narrowing and those who believe in the opposite. The perception-reality gap may be attributed to a number of factors. Since the 'birth' of the digital divide, the phenomenon has largely been perceived as an ICT-dependent artefact. Consequently, indices for measuring the digital divide have tended to be techno-centric. Moreover the digital divide is, by and large, a Eurocentric creation, and in its application (as a construct) has hardly taken into account the peculiarities of the developing world. Woherem (1993) commenting on the use of ICTs in developing countries, pointed out that unless ICTs were integrated into the cultural milieu of communities, people would stand fewer chances of benefiting and therefore accepting such technologies. He observed that the software applications used in developing countries are based on western models and do not take into account local cultural sensitivities. Moreover, much of the technology used in the developing world was transplanted without any re-engineering to suit local conditions.

The digital divide is now increasingly evolving into (or exists side by side with) the so-called new digital divide, alienating further those who have traditionally been removed from digital inclusion. For example, the EU conceived the idea of e-inclusion in the context of e-government to refer to the employment of modern ICT technologies to address issues of the access-divide and promote opportunities for the economic and social empowerment of all citizens (United Nations, 2005). E-inclusion envisages a future in which all people have access to social and economic opportunities and can use technology. Measures of the digital divide have also tended to aggregate or disaggregate technologies when determining the level of the digital divide. The fallacy of (des) aggregating technologies is that each technology is unique in its diffusion, penetration and use. When findings from disaggregated measures are generalised across different technologies, they may not give a real picture of the depth and breadth of the digital divide. For example, the fact that Africa has the highest mobile phone penetration in the world does not mean that the digital divide is effectively narrowing between Africa and the developed world. This is primarily because PC penetration, Internet penetration, access to broadband, use of existing technologies, etc., remain low on the continent. Sciadas (2002) observes that there are old and new ICTs, and each is distinct in its attributes, functionality and numerous other characteristics, including pricing. There is therefore no reason to lump different ICTs in one group and expect similar patterns in their penetration, either across groups or over time.

The digital divide is also assessed based on the increasing number of 'digital natives' compared to the number of 'digital immigrants'. 'Digital natives' refers to individuals who are born in a technologically advanced society, and who are therefore raised with technology, play as youngsters with technology, learn with technology and expect to spend their productive years working with technology (Broere and Kruger, 2007). 'Digital immigrants', on the other hand, are those who were already socialized in pre-digital ways before digital technology arrived on the scene (Prensky, 2001).

Bridging the digital divide is commonly but fallaciously perceived as bringing on board the people (often constituting the majority) who remain excluded from access to technology, especially those living in the rural areas of most developing countries. However, some of the so-called knowledge centres situated in rural areas in India (new India) - meant well in principle to bridge the urban/rural divide - found that some of the residents living next to these centers, either through omission or commission, were not keen to use them (The Economist Newspaper and the Economist Group, 2005). This contrasts sharply with the optimism of the 1990s that rural ICTs would allow previously marginalized communities to leapfrog development, information societies and a host of other electronic age applications. This situation calls for research into the unexamined rural-rural digital divide. Crump and Mclroy (2003), discussing a community-based project in Wellington, New Zealand, wonder why when computing is available in a socially situated, convenient environment, at no cost, some people choose not to compute. Similarly, Lenhart, Horrigan, Rainie, Allen, Boyce, and Madden, et al. (2003), also in a research project in Wellington on economic and social inclusion, noted that the polarization of the "haves" and "have nots" based on the premise of physical access to ICTs is flawed. This is because not all "have nots" necessarily want to be "haves", and not everybody views their engagement (even hypothetically) with ICTs as a positive force that would improve the quality of their lives.

In Northern Ireland (UK), the free provision of computers, fast Internet access and a website to residents/businesses of Ennis in 1997 to enhance ICT uptake and modernise society was met with limited success. Three years later, the initiative had little to show for it, because technology had been thrust into people's hands without their participation and with little preparation. Training programs had been run, but they were not sufficiently accompanied by programs that demonstrated why people should use the new technology in the first place. Goode (2001) posits that the digital divide should really be looked at from the perspective of meaningful access. In order to use computers as tools for education, they must have relevant applications in the user's life.

Bill Gates observed that community centres or similar ventures are distractions from real problems of development, and noted that 99% of the benefits of having access to a PC come when the person who sits down to use it has been provided with reasonable levels of healthcare and education (The Economist Newspaper and The Economist Group, 2005). Elsewhere, an e-readiness assessment study in Botswana in 2005 revealed that rural communities preferred interventions that would enhance access to anti-retroviral drugs, information about where they could fetch good market prices, and information on how to track lost cattle, as opposed to interventions that would gain them access to the Internet or computers (Maitlamo, 2005). Similarly, in an e-government presentation paper in South Africa, people were reported to be more preoccupied with issues of security, shelter, access to clean water, electricity, and access to health care facilities than with Internet connectivity (Geness, 2004). Warschauer (2002) noted that a social inclusion framework redirects the focus from providing access to technology to the effective integration of ICT into communities and institutions for social development. This kind of integration can only be achieved by paying attention to the holistic array of physical, digital, human and social resources available to or within local communities.

Although the debate on whether the digital divide is narrowing or widening is healthy, it fails to recognise that the digital divide is not about a single technology, but about a number of technologies (e.g. PCs, mobile phones, Internet, etc), and is driven by a complex set of factors that exist beyond wires. Moreover, the digital divide is not centred on affordability because research in extant literature has shown that there are people with ready access to information technologies in both the developed and developing worlds who are reluctant to use ICTs such as the Internet (Green, 2000; Warschauer, 2002). The digital divide has often been perceived as inequitable access to ICTs such as PCs, Internet, telephones, cable and other Internet-related technologies by individuals or groups of people in a country or across countries (Spectar, 2000). This perception of the digital divide was apparent at the World Summit on Information Society in 2003, when poorer countries, particularly those from Africa, lobbied successfully for the establishment of a "Digital Solidarity Fund" to help finance "infrastructure" that is "very much needed" to close the perceived technology gap. African member states argued that one of the main problems affecting access to ICT on their continent was lack of adequate requisite infrastructure, including telephone access, mass media and other forms of communication systems (PANOS, 2004).

Definitions of the digital divide that emphasize technologies may be viewed as the 'first generation' type. However, recent literature expands on the scope of the term, thus pushing the phenomenon to the next (second generation) level. For example, the International Telecommunications Union (2002) observes that the so-called "new" or "quality" digital divide cannot be attributed to the lack of equipment or connections alone; rather the nature of the phenomenon is changing from "basic to advanced communications (i.e. from copper-based access to fibre optic access) and from quantity to quality". Norris (2001) supports this view and suggests that the digital divide requires us to look beyond the issue of access to technology. The digital divide is therefore a multidimensional phenomenon encompassing three distinct aspects, namely: the global divide, which refers to differences in Internet access across industrialized and developing societies; the social divide, which is concerned with the gap between the information rich and information poor in each nation; and the democratic divide, which signifies the difference between those who do and those who do not use the panoply of digital resources at their disposal to engage and participate in public life.

Despite the increasing recognition that the digital divide goes beyond access to wires, efforts to bridge the phenomenon in the past have been driven by the first and second generation definitions. Examples include the Digital Opportunity Task Force (2002); National Information Infrastructure (NII) and Global Information Infrastructure (GII) projects in the US (Miranda, 2006); African Information Society Initiative (AISI) (Amoako, 1996); New Partnership for Africa's Development e-school programme (Association for Progressive Communications, 2005); East African Submarine Cable System (EASSy); SAT-3/WASC or South Atlantic 3/West Africa Submarine Cable; Common Market for East and Southern Africa (COMESA) Telecommunications project (COMTEL); and Kenya's government-sponsored undersea fibre optic cable, known as The East Africa Marine System [TEAMS] (Morris, 2007).

The departure from the first and second generation definitions of the digital divide and the ushering in of what could be considered the third generation type may have emerged during the World Summit on Information Society (International Telecommunication Union, 2003) held in Geneva, Switzerland, in 2003. In its Declaration of Principles, the summit declared the common desire and commitment to build a people-centred, inclusive and development-oriented information society, where everyone can create, access, utilize and share information and knowledge; enabling individuals and communities to achieve their full potential by promoting sustainable development and improving their quality of life. In this declaration, WSIS went beyond access and explicitly/implicitly emphasized "access to information and knowledge, capacity building, building confidence and security in the use of ICTs; enhancing cultural and linguistic diversity, access to local content; and ethical dimensions for an inclusive information society".

Like WSIS, recently developed indices for measuring digital inclusion, such as the digital opportunity index, information society index and e-government index, demonstrate how complex the digital divide phenomenon really is. For example, the Digital Opportunity Index (DOI) measures and evaluates the opportunities, infrastructure and utilization of ICTs. DOI monitors the growth of mobile communications, as well as more recent technologies such as broadband and mobile Internet access. It also looks at the falling prices of broadband and increasing broadband speeds (World Information Society Report, 2006).

Correspondingly, the Information Society Index examines how nations are positioning (or have positioned) themselves to compete in the global information economy. The index is based on variables such as: IT spending as a percentage of GDP, software spending, IT services' spending, PC penetration, Internet users, home Internet users, mobile Internet users, e-Commerce spending, broadband households, wireless subscribers, secondary education levels, tertiary education levels, civil liberties, and government corruption (International Data Corporation, 2004). Likewise, the e-Government Index measures the extent of the digital or technological divide between countries and focuses on current practices in government with regard to digital governance (delivery of public services), digital democracy (citizen participation in government), security, usability, content of websites and the type of online services on offer (Holzer and Kim, 2005). WSIS's vision and its Declaration of Principles, together with the DOI, ISI, E-readiness and E-government indices, provide the framework and motivation for a new model that deconstructs the digital divide and attempts to explain how it develops.

DIGITAL DIVIDE AND THE ECONOMIC DEVELOPMENT NEXUS

Governments the world over have been preoccupied with bridging the digital divide, as they now believe that there is a direct correlation between digitization and economic development (Dutta et al., 2004; World Economic Forum, 2003). Zaidi (2005) defines economic development as growth in the GDP accompanied by relevant social and institutional changes by which that growth can be sustained. These changes include reduced levels of absolute poverty, a better quality of life, high literacy levels, improved labour productivity, sophisticated techniques of production, development of physical and commercial infrastructure, higher savings, increased employment opportunities, a positive attitude towards life and work, and a stable political system. Efforts aimed at bridging the digital divide must therefore first and foremost uplift the socio-economic status of the people. Fink and Kenny (2004) point out that in many developing countries, people face far more critical challenges (such as lack of access to water, food, medical treatment and education) than the lack of Internet access. They conclude that international aid should be directed towards meeting basic needs rather than to narrowing the digital divide.

The digital divide is closely linked to socio-economic development; consequently efforts to address it should simultaneously address issues of poverty, especially in Africa where it seems to be on the increase. Dlamini (2004) argues that poverty in Africa appears to be on the rise, seeing as Africa's GDP in 2004 accounted for only 1.5% of the global GDP. Similarly, Africa's share of trade during the same year was estimated at 2.1%. The correlation between high poverty levels and increased digital inequality has been established in literature. For example, the United Nations Conference on Trade and Development (UNCTAD, 2006) found that a person in a high-income country is over 22 times more likely to be an Internet user than an individual in a low-income country. Similarly, secure Internet servers - a rough indicator of electronic commerce - are over 100 times more common in high-income than in low-income countries. In high-income countries, mobile phones are 29 times more prevalent and mainline penetration is 21 times that of low-income countries. The cost of Internet access in a low-income country is also approximately 150 times the cost of a comparable service in a high-income country. UNCTAD concludes that growth in the use of technology in business, schools and homes could raise the standards of living and help people prosper.

BBC News (2008), in an article entitled 'mobiles narrow digital divisions', notes that mobile phone and net access are helping narrow the gulf between rich and poor nations. The efficiencies that these technologies bring have boosted development in poorer countries. The report estimates that mobile phone users in developing nations now make up over 58% of the total handset subscribers worldwide. In Africa, which has experienced the greatest increase in mobile phone subscribers and penetration, this technology has been said to have the potential to improve the economic life of the continent's population as a whole. The report notes that among rural communities in Uganda and small vendors in South Africa, Senegal and Kenya; mobile phones are helping traders get better prices and allowing traders to sell goods faster, thus ensuring that less is wasted.

Peters (2003) also affirms that installing computers and connections in underdeveloped communities is necessary to put ICT to use for socio-economic development. The African Information Society Initiative (AISI), at its 21st meeting in May 1995 during the Economic Commission for Africa (ECA) Conference of Ministers of Social and Economic Development and Planning, set to build Africa's first information super highway (a process that is still ongoing in most countries), which would utilize ICTs to accelerate the socio-economic development of Africa and its people (Amoako, 1996).

CAUSES OF THE DIGITAL DIVIDE

The causes of the digital divide in general are well documented in existing literature (Mutula, 2006). However, there are other emerging causes of the divide that have not gained much ground in digital divide literature. For example, Waters (2008) notes that in Africa, 70% of Internet traffic is routed through Europe and North America, thus driving up costs for businesses and consumers. This means that a great number of businesses are unable to get or maintain connectivity. In addition, Internet governance is skewed in favour of the developed world. For example, the English language has often been blamed for disrupting communication across the developed and developing worlds. Similarly, the requirement that all domain names must be entered in ASCII characters in order to support the Latin alphabet means that diacritical marks, and Asian or other international characters, are not supported. Consequently, the exclusion of a number of other languages from domain names limits Internet access. Users who are not familiar with English may encounter problems when accessing English language URLs. Additionally, the lack of foreign-script support makes it difficulty for indigenous businesses and entities to be represented on the Internet.

The digital divide is also exacerbated by intellectual property regimes. For example, conventional practice with regard to copyright law states that the duration of copyright extends from the life of the author to fifty years after his/her death. After 50 years, copyrighted materials move into the public domain and can be used freely without any charge. The US amended its law (Copyright Extension Act of 1998), increasing the duration of copyright to 70 years (House and Senate, 1998). The US has also tried to apply pressure on national governments, including those in Africa, to change the conventional copyright law (Kiggundu, 2007). This has implications in terms of the capacity of impoverished African countries to afford such material for sustained periods of time. At the same time, it gives the US leverage to allow its powerful multinational corporations to reap continuing profits at the expense of poorer countries.

Within electronic environments, large amounts of information are generated and used in digital formats in the form of e-books, databases, e-journals, etc. Most of the content providers of digital material are in the developed world, and whenever they enter into negotiations with information providers in developing countries, they insist on the inclusion in the agreements of contractual terms. These override traditional exceptions to copyright as contained in national legislations, such as fair use and fair dealing. It is therefore difficult for information providers to freely avail such information for academic use without breaking license agreements, further alienating developing countries from mainstream information.

The digital divide is also exacerbated by the high cost of digital information. The high cost of access to external databases and for the procurement of digital information makes it increasingly difficult for libraries in the developing world to subscribe to new journals and books, or even maintain existing subscriptions. Libraries the world over have well established traditions of archiving hard copies of all the materials that they acquire, be they journals, books, newspapers or pamphlets. But it is increasingly becoming difficult for libraries to continue with this long established practice because digital content providers are insisting that once subscription is terminated (due to non-payment, for example), access to entire databases is also lost. Unlike print-based sources, there can be no copies made or put aside once access to an electronic resource is terminated.

Efforts to bridge the digital divide in developing countries are also in part hampered by the fact that most countries have not implemented broadband technologies. For those that have made efforts to provide access to broadband, the cost remains prohibitive (Internet World Statistics, 2007).

INTERVENTIONS TO BRIDGE THE DIGITAL DIVIDE

Countries in the developing world need to take prudent measures to bridge the digital divide in their midst. Efforts should be made to establish Internet exchange points that would obviate the need for traffic within the continent to be routed through Europe or North America. The long term strategy for bridging the digital divide should include the implementation of ICT education in the curriculum throughout schools. Goode (2001:19) posits that if we do not educate our students on the appropriate use of digital technologies, we are not preparing them for full participation in our democracy (and by extension, the information society).

Ray (2007) observes that mobile number portability in the UK has been a critical intervention in stemming the digital divide in the region. Number portability refers to keeping your phone number when you change networks. Other measures have included providing incentives that can help reduce the income tax levied on IT companies; VAT exemption on locally purchased ICTs; improving or enhancing the energy grid; reducing the costs of bandwidth available to public universities through subsidies; and capacity building in the area of ICT applications. Mobile technology can potentially minimize the digital divide, especially in developing countries. WiMAX, or "Wireless Maximised" technology, allows fixed-line telecommunications systems to utilize mobile technology to carry voice and data beyond the farthest reach of land lines and thereby deliver a wireless "final mile" that can reach up to 30 miles from copper lines in any direction. This could potentially extend Internet access to most developing populations and save at least some capital investment in new land-lines. Similarly, VSAT--Very Small Aperture Terminal--is a direct satellite-to-end-user technology that skips terrestrial wiring completely and thus presents an alternative to land-line communication.

Policies that promote open Internet access are also needed to stem the digital divide between developing countries and the developed world. In the US for example, high speed Internet policies have been implemented to enable government to promote a more open Internet. High speed Internet is necessary for economic growth, employment, telemedicine, education, e-government, and public safety (Communication Workers of America, 2006). Developing countries could also benefit from open access initiatives in their attempt to bridge the digital divide. In order for this goal to be realized, authors, publishers, libraries and other stakeholders should work together to develop suitable business models. Developing countries may also exploit the opportunities offered by blogging to create local content. Blogs give authors ownership of the medium of communication, and allow users to instantly publish text or graphics on the web without the need for sophisticated technical knowledge. Blogs have great potential in improving interaction online because they can be easily integrated with other learning technologies. In addition, they can effectively be used by all age groups and both genders as a medium for learning (Huffaker, 2004).

Developing countries, like the European Union member states (European Commission, 2005), should develop policies that cater for vulnerable groups in society. National legislation and regulatory frameworks are needed that ban the sale of inaccessible technology products, while enhancing the growth of assistive technology. National strategies should emphasize the interoperability of products, education, universal service policies for electronic communications, affordable pricing of networks, and interactive content. Effective management of copyright may also assist developing countries in bridging the digital divide. Kiggundu (2007) suggests that developing countries should take a firm stand with regard to provisions in the Copyright and Neighbouring Rights Act of 2000, which expressly prohibits the inclusion of restrictive contractual terms in any contract entered into with information providers, and also by advocating in WIPO and WTO for such restrictive terms to be prohibited worldwide.

The world over, most countries are moving government services online. E-government infrastructure provides an effective means to expand the technological capabilities of citizens and businesses for participation in the governance process and to meet the growing demands of citizens and businesses. E-government is also an important infrastructure for enhancing local content development. It therefore provides opportunities for bridging the digital divide, especially between urban and rural areas.

The developing world should embrace open source products to obviate the need for translators from the different application formats and allow the reuse of existing work across all projects easily. Allowing people to use the computer in their mother tongue would stimulate pride in their language, and learning something in one's mother tongue is also naturally easier.

BENEFITS OF BRIDGING THE DIGITAL DIVIDE

In the 1990s, governments the world over initiated efforts to bridge the digital divide in order to achieve digital dividends for their citizens in the social, economic and political spheres. Among the digital dividends envisaged were improved service delivery, the prudent management of public resources, and overall good governance. Since 2000, governments have been preoccupied with putting in place strategies to meet the Millennium Development Goals (MDGs) by the year 2015. Universal access to digital technologies is seen as an important catalyst in this endeavour.

The World Summit on Information Society (WSIS, 2003) noted that the digital revolution, fired by the engines of information and communication technologies, has brought with it new fundamental ways of creating knowledge, educating people and disseminating information; conducting economic and business practices; running government; engaging in politics; providing speedy delivery of humanitarian aid and healthcare; and improving the living standards of millions of people around the world. Bridging the digital divide therefore creates opportunities for previously marginalized groups. Such opportunities include access to education, equal employment opportunities, and medical care.

The United States government integrated ICT into its policy documents primarily to achieve economic development and increased productivity. The US advanced this idea by advocating Global Information Infrastructure (GII) through the World Trade Organization and the North American Free Trade Agreement (NAFTA). At the 1994 ITU international conference in Buenos Aires, Albert Gore noted that the NII and GII would lead to sustainable economic progress and improved health care, among other benefits. Moreover, GII would lead to a global free market and global decentralized democracies, more freedom for individuals, and more choices (Miranda, 2006).

The world over, countries that are highly digitally proficient are also performing well in terms of economic development. For example, countries such as Switzerland, Finland, Sweden, Denmark, Singapore, the United States, Japan, Germany, the Netherlands and the United Kingdom are among the world's top ten performing economies, according to *The Global Competitiveness Report, 2006-2007* (World Economic Forum, 2006). These countries are also leaders in e-governance systems and score highly on the digital opportunity index, e-readiness rankings, information society indices and e-government indices.

CONCLUSION

Bridging the digital divide has become a preoccupation of many countries around the world, but emphasis has tended to focus on enhancing access to technologies. Consequently, there is a misconception that increasing access is panacea to narrowing the digital divide. However, as demonstrated in this chapter, there is overwhelming evidence to the contrary - that the digital divide between the developed and developing world seems to be widening despite claims that it is narrowing. Until the digital divide is properly deconstructed and understood, efforts to bridge it will remain futile. Moreover, as various measures such as e-readiness, information society, digital opportunity, and e-government indices have shown, bridging the digital divide is positively correlated to economic development. Consequently, efforts to bridge the digital divide should first and foremost address issues of economic development, especially poverty reduction in society. Various efforts to bridge the digital divide should also be followed by impact assessments to determine whether the desired outcomes have been achieved. Impact studies would provide an indication of where changes in strategy and more investment with regard to infrastructure development and policy are needed.

REFERENCES

Amoako, K. Y. (1996, May 15). Keynote address. *Conference on Empowering Communities in the Information Society,* South Africa. Retrieved on August 28, 2004, from http://www.uneca.org/eca_resources/Speeches/amoako/96_97/empower2.htm

Arbore, A., and A. Ordaini (2006). Broadband divide among SMEs: The role of size, location, and outsourcing strategies. International Small Business Journal, 24(1), 83–99. doi:10.1177/0266242606059781

Arendt, L. (2008). Barriers to ICT adoption in SMEs: How to bridge the digital divide? Journal of Systems and Information, 10(2), 93–108. doi:10.1108/13287260810897738

Association for Progressive Communications. (2005). *Kenya: 6 schools to gain from NEPAD ICT programme.* Retrieved on April 20, 2007, from http://rights.apc.org/africa/index.shtml?apc=21878ne_1&x=554614

Atkins, T. (2005). Digital divide narrowing fast, World Bank says. *BBC News, 13*(52).

BBC News, (2007). Mobiles narrow digital divisions. Retrieved on February 8, 2008, from http://news.bbc.co.uk/2/hi/technology/7232264.stm

Birdsall, W. F. (1997). The ideology of information technology. Queen's Quarterly, 104, 287–299.

Broere, I., & Kruger, M. (2007, September 5-7). *Engaging digital natives: Accelerating change towards the future.* Paper presented at the 9th Annual Conference on WWW Applications, Johannesburg, South Africa. South Africa: University of Johannesburg.

Citizens Online. (2007). *Evidence of widening digital divide*. Retrieved on September 13, 2007, from http://www.citizensonline.org.uk/conline/news/display?contentId=4686

Civilrights.org (2005). *New research shows persistence of digital divide*. Retrieved on August 5, 2007, from http://www.civilrights.org/issues/communication/remote-page.jsp?itemID=28373283.

Communication Workers of America. (2006). *Bridging new digital divide: Need for national policy*. AFL-CIO, CLC: Communications Workers of America.

Crump, B., & Mclroy, A. (2003). *The digital divide: Why the don't-wants-tos 'won't compute: Lessons from a New Zealand ICT project*. Wellington: First Monday.

Digital Opportunity Task Force. (2002). *Digital opportunities for all: Meeting the challenge, report of the digital opportunity task force including the Genoa plan of action*. Retrieved on July 13, 2002, from http://www.dotforce.org/reports/DOT_Force_Report_V_5.0h.html

Dlamini, K. (2004). Build Africa into brand worth buying. *Sunday Times Business Times, 13.*

Dutta, S., Lanvin, B., & Paua, F. (2004). *The global information technology report: Towards an equitable information society*. Oxford: Oxford University Press.

Economic Commission for Africa. (2006). *Per capita electricity use in Africa is less than 2 percent, says new report by ECA's NEPAD unit*. Retrieved on April 2, 2007, from http://www.uneca.org/nepad/Story070326.htm

The Economist Newspaper and the Economist Group(2005). *Behind the digital divide*. Retrieved on April 10, 2007, from http://www.economist.com/printedition/PrinterFriendly.cfm?Story_ID=3714058

European Commission. (2005, November 24-25). Transforming public services. *Report of the Ministerial E-Government Conference*, Manchester, UK. Retrieved on May 10, 2007, from http://archive.cabinetoffice.gov.uk/e-government/conference

Fink, C., & Kenny, C. (2004). Canyon or mirage? *Economist, 370*(8359). Retrieved on March 14, 2008, from http://web37.epnet.com/citation.asp?tb=1&_ug=sid+D9D7A499%2DB848%2D4902%2D86CB%2DE18C165E13AB%40sessionmgr3+dbs+aph+cp+1+9D8B&_us=hs+True+cst+0%3B2+or+Date+ss+SO+sm+KS+sl+0+dstb+KS+ri+KAAACBBD00032457+0E2B&_uso=hd+False+tg%5B0+%2D+st%5B0+%2DKenny++and++Fink++and++digital++divide+db%5B0+%2Daph+op%5B0+%2D+65C6&cf=1&fn=1&rn=1

Geness, S. (2004, April 14-16). *E-Government, the South African experience*. Paper presented at SADC E-Government Workshop, Gaborone.

Goode, J. (2001). *Employing technology in education*. Retrieved on August 5, 2007, from http://tcla.gseis.ucla.edu/divide/teachers/goode.html

Green, L. C. (2000). *Bridging the digital divide in our schools–achieving technology equity for all students: Education market research*. New York: Education Market Research.

Holzer, M., & Kim, S. T. (2005). *Digital governance in municipalities worldwide: A longitudinal assessment of municipal Web sites across the world*. Retrieved on February 15, 2007, from http://unpan1.un.org/intradoc/groups/public/documents/ASPA/UNPAN022839.pdf

House and Senate. (1998). Copyright extension act. *Tech Law Journal*. Retrieved on September 12, 2007, from http://www.techlawjournal.com/courts/eldritch/pl105-298.htm

Huffaker, D. (2004). *Educated blogger: Using weblogs to promote literacy in classroom*. Retrieved on September 26, 2006, from http://www.firstmonday.org/issues/issue9_6/huffaker/index.html

International Data Corporation. (2004). *Information society index: Middle East and Africa*. Retrieved on February 22, 2006, from http://www.idc.com/groups/isi/main.html

International Telecommunications Union. (2002). *World telecommunication development report: Reinventing telecom, 6th ed.* Geneva: International Telecommunication Union.

International Telecommunications Union. (2003). World summit on the information society. Geneva 2003 Tunis 2005. Retrieved on March 13, 2007, from http://www.itu.int/wsis/docs/geneva/official/dop.html

International Telecommunication Union. (2005). *ITU and orbicom publish ICT opportunity index: A step towards the implementation of the WSIS' plan of action*. Retrieved on April 17, 2007, from http://www.itu.int/ITU-D/ict/publications/dd/summary.html

Internet World Statistics (2007). *Internet usage statistics: World Internet users and population stats*. Retrieved on September 12, 2007, from http://www.internetworldstats.com/stats.htm

Kiggundu, J. (2007, April 14). University education and intellectual property in the digital era: Whither Botswana? *Seminal Paper*. Gaborone: University of Botswana.

Lenhart, A., Horrigan, J., Rainie, L., Allen, K., Boyce, A., & Madden, M., et al. (2007, June 25-27). Current status of SMEs in Taiwan. *12th APEC Women Leaders Network (WLN) Meeting,* Port Douglas, Australia.

Maitlamo. (2005). *ICT sector survey in Botswana. Report of national ICT consultancy for Botswana.* Gaborone: Consulting and Audit Canada.

Miranda, A. D. (2006). Technological determinism and ideology: Questioning the 'information' society and the 'digital divide.' Retrieved on April 2, 2007, from http://www.ces.uc.pt/bss/documentos/2006_11_13_alvaro_de_miranda.pdf

Monge-González, R., Alfaro-Azofeifa, C., & Alfaro-Chamberlain, J. I. (2005). *TICs EN LAS PYMES DE CENTROAMÉRICA: Impacto de la adopción de las tecnologías de la información y la comunicación en el desempeño de las empresas*. IDRC.

Morris, A. (2007). *SatCom Africa celebrates 10 years of business excellence*. Retrieved on April 9, 2007, from http://www.satcomafrica.com/

Mutula, S. M. (2006). Freedom of information in the SADC region: Implications for development and human rights. Library Review, 55(7), 440–449. doi:10.1108/00242530610682155

Naidoo, S. (2007). Telkom has lost its head. *Business Times*. Retrieved on April 18, 2007, from http://www.mybroadband.co.za/nephp/?m=show&id=6099

Norris, P. (2001). *Digital divide: Civic engagement, information poverty, and the Internet worldwide.* Cambridge: Cambridge University Press.

Nyasato, R., & Kathuri, B. (2007). High phone charges hamper region's growth, says World Bank. *The Standard.* Retrieved on April 10, 2007, from http://www.eastandard.net/hm_news/news.php?articleid=1143967136

PANOS. (2004). *Completing the revolution: The challenge of rural telephony in Africa. PANOS report no. 48*. London: The PANOS Institute.

Peters, T. (2003). Bridging the digital divide-global issues. *An Electronic Journal of the U.S. Department of State, 8*(3). Retrieved on April 17, 2007, from http://usinfo.state.gov/journals/itgic/1103/ijge/gj08.htm

Prensky, M. (2001). *Digital natives, digital immigrants*. Retrieved on June 1, 2005, from http://www.marcprensky.com/writing/Prensky%20-%20Digital %20 Natives,%20Digital%20Immigrants%20-%20Part1.pdf

Ray, B. (2007). *Moving mobile numbers should be instant*. Retrieved on August 5, 2007, from http://www.theregister.co.uk/2007/07/18/number_portability/

Riccardini, F. (2003, September). The information society: Measuring the SMEs and the digital divide. *UNCTAD Expert Meeting-Measuring Electronic Commerce*, Geneva.

Sciadas, G. (2002). *Unveiling the digital divide*. Ottawa, Ontario: Science, Innovation, and Electronic Information Division.

Spectar, J. (2000). Bridging the global divide: Frameworks for access and the world wireless Web. *North Carolina Journal of International Law and Commercial Regulation, 26*(57).

United Nations. (2005). *Global e-government readiness report*. Retrieved on April 17, 2007, from http//: unpan1.un.org

United Nations Conference on Trade and Development UNCTAD. (2006). *The digital divide report: ICT diffusion index, 2005*. Retrieved on September 23, 2007, from http://www.unctad.org/templates/webflyer.asp?docid=6994&intItemID=2068&lang=1&mode=downloads

Warschauer, M. (2002). Reconceptualizing the digital divide. *First Monday, 7*(7). Retrieved on March 14, 2008, from http://firstmonday.org/issues/issue7_7/warschauer/index.html

Wielicki, T. (2008). ICT training–key to closing digital divide among businesses: Case of SMEs in central California. *The International Journal of Learning*. Retrieved on November 22, 2008, from http://ijl.cgpublisher.com/product/pub.30/prod.1428

Woherem, E. E. (1993). *Information technology in Africa: Challenges and Opportunities*. Nairobi: African Centre for Technology Studies.

World Economic Forum. (2003). *The global information technology report: Towards an equitable information society*. Johannesburg: WEF.

World Economic Forum. (2006). Global competitiveness report 2006-2007. Retrieved on April 18, 2007, from http://www.weforum.org/en/initiatives/gcp/Global%20Competitiveness%20Report/index.htm

World Information Society Report. (2006). *Digital opportunity index 2005*. Retrieved on February 13, 2007, from http://www.itu.int/osg/spu/publications/worldinformationsociety/2006/World.pdf

WSIS. (2003). *World summit on the information society: Draft declaration of principles*. Geneva: WSIS.

Zaidi, H. H. (2005). *The meaning of economic development*. Retrieved on April 11, 2007, from http://www.dawn.com/2005/01/17/ebr2.htm

Section 3
E-Readiness Assessments

This section consists of Chapters 6, 7, and 8 (E-Readiness and Its Assessment, E-Readiness Assessment Methods and Tools, and E-Readiness of SMEs, respectively). Chapter 6 covers the international status of e-readiness across different jurisdictions (Europe, North America, Asia-Pacific, Middle East, and Africa), its implications for businesses (especially SMEs), and the importance of e-readiness assessments for businesses and governments. The purpose of Chapter 6 is to demonstrate how the level of e-readiness affects the competitiveness of businesses, and also to highlight managerial and policy implications.

Chapter 7 covers measures of e-readiness, macro and micro e-readiness assessment tools (including the tools' names, the developers of the tools, intended applications of the tools, the countries in which the tools have been used, and an overall analysis of the tools), strengths and weaknesses of e-readiness assessment tools, and choosing the right assessment tool. The chapter aims to highlight the different types of tools available for use by SMEs or governments that wish to measure their e-readiness and gauge their position in the global digital economy. It also illustrates that while e-readiness tools for SMEs are limited, macro e-readiness tools that are used for measuring national e-readiness may be applied with adaptations.

Chapter 8 (E-Readiness of SMEs) examines how the e-readiness of SMEs is assessed. It covers the strengths of SMEs over large scale enterprises in the digital economy, the types of transactions that SMEs undertake online, the e-readiness adoption process among SMEs, international e-readiness status of SMEs (in North America Europe, the Middle East, Asia-Pacific, the Caribbean, and Africa), and how to get SMEs to reach an e-readiness status. The purpose of this chapter is to enable policy makers and managers of SMEs to compare the relative positions of SMEs in the global business environment so that they may put in place requisite policies and institutional infrastructure to enhance SMEs competitiveness in the digital economy.

Chapter 6
E-Readiness and Its Assessment

INTRODUCTION

The level of e-readiness of the digital economy is expected to vary across different jurisdictions. The e-readiness ranking of nations, which has been a regular undertaking by various agencies since 2000, demonstrates that the world is, indeed, moving steadily into a new digital era, and that by extension, businesses are increasingly operating in the digital economy. The 2008 e-readiness ranking of nations and similar previous rankings suggest that collectively, the world is steadily moving up the e-readiness charts; for example, average e-readiness rose by 0.15 to 6.39 in the 2008 rankings, up from 6.24 in the previous year, with the United States being the 2008 global e-readiness leader (with 8.95), followed closely by Hong Kong, which advanced up by two places (The EIU/IBM Business Institute for Business Value, 2008).

DOI: 10.4018/978-1-60566-420-0.ch006

Countries that are leaders in e-readiness rankings, such as the US, Hong Kong, the Netherlands, Australia, Denmark, Switzerland, etc; have made significant improvements in the expansion of connectivity in broadband and WiFi, and have also enhanced the security of Internet connections. The 2008 e-readiness report reads that "...the gap between the "haves" and "have-nots" in [the e-readiness] rankings narrowed again in 2008, a hopeful indication of a contraction in the digital divide between developed and developing countries and movement into the digital economy". The move of the world towards the digital economy is also reflected in the increasing penetration of mobile phones in regions that were traditionally marginalized in telecommunications ubiquity, especially in parts of the developing world such as Africa (The EIU/IBM Business Institute for Business Value, 2008).

The 2006 e-readiness report noted the following: "The world in 2006 is ever more "e-ready"—there are over 1bn Internet users and 2bn mobile-phone users worldwide, and most countries continue to make steady progress in most qualitative indicators of technology-related development" (The Economist Intelligence Unit and IBM Corporation, 2004. p. 3). Virtually all the countries included in 2006's e-readiness rankings improved their scores over the next year. Moreover, both in relative and absolute terms, the improvement was greater in the lower tiers of the rankings than at the top. As a result, the distance separating the best from the rest was narrowed. Moreover, poorer countries were moving up the e-readiness scale faster than richer ones. Without a doubt, the commercialisation of the Internet and World Wide Web (WWW) has made significant contribution to the increasing use of this medium, especially with regard to communications, government and e-commerce.

Internationally, the phenomenon of e-readiness is regularly assessed by various agencies in government, academia, business and the NGO sector, and the results are then either published on an annual basis, or two years thereafter. For example, in 2004, The Economist Intelligence Unit and IBM Corporation (2004, p. 3) ranked 64 countries according to their e-readiness using both quantitative and qualitative indicators, including technological infrastructure, the business environment, degree to which e-business was being adopted by consumers and companies, social and cultural conditions that influenced Internet usage, and the availability of services to support e-business. The results are shown in Table 1, Table 2, and Table 3.

E-Readiness Status in Europe

With the exception of Greece (27th), most countries in Western Europe made it into the top 25 list for that year, as shown in Table 1.

The countries in the top 25 tier were characterised by thriving economies, sophisticated infrastructure, coordination between government and the private sector, and

Table 1. E-Readiness status in Europe

Country	Rank
Denmark	1
United Kingdom	2
Sweden	3
Norway	4
Finland	5
Netherlands	8
Switzerland	10
Germany	13
Ireland	16
Belgium	17
France	18

the increasing growth of the Internet among businesses and consumers (Economist Intelligence Unit and IBM Corporation, 2004:8). In Eastern Europe, gradual market liberalisation had transformed the telecommunications and IT sectors. As monopolies were privatised and competition was let in, services improved and prices fell. Countries such as Estonia, Latvia, Poland, Lithuania, Slovakia, Cyprus and Malta had decent infrastructure and e-business environments. The top ranked countries in Eastern Europe in the 2004 survey are reflected in Table 2.

The e-readiness status in Europe was boosted by the European Union, which launched a concerted and widespread campaign for the proliferation of the Internet

Table 2. E-Readiness status in Eastern Europe

Country	Rank
Estonia	26
Czech Republic	27
Hungary	30
Slovenia	31
Latvia	34
Poland	36
Lithuania	38
Slovakia	39
Bulgaria	50

Table 3. E-readiness status in North and South America

Country	Rank
Canada	10
Chile	28
Brazil	36
Argentina	37
Mexico	39
Colombia	41
Venezuela	44
Peru	47
Ecuador	56

throughout the continent; the promotion of e-commerce in both the public and the private sector; the widespread adoption of the electronic delivery of government services; and the provision of online information. The concept behind this strategic thrust was encapsulated in the European Commission's document entitled '*e-Europe - An Information Society for All*'. The EU recognised that e-commerce impacted positively on business performance and national economic growth. Consequently, non-participation in this revolution implied the risk of falling behind in the economic front, with all the implications that this would entail (Borg, 2004:6).

The EU, through its e-Europe 2002 ICT Action Plan, delivered major changes and increased the number of its citizens and businesses connected to the Internet. Most of the EU member states reshaped the regulatory environment for communications networks and services and e-commerce, and opened doors to new generations of mobile and multimedia services (Commission of the European Communities, 2002:3). In all the European member states, Internet penetration has doubled, telecommunications frameworks are largely in place, Internet prices have fallen, and e-commerce legal frameworks have been put in place.

E-Readiness Status in North and South America

In North America, the United States and Canada were reported by the Economist Intelligence Unit to have achieved high rates of e-readiness in 2004, with absorption rates of approximately 70%, positioning them amongst the world's leaders. In both countries, telecommunications infrastructure was strong, and businesses and consumers had swiftly and effectively integrated online processes into their daily activities. In terms of e-readiness ranking within the Americas in 2004, the US led, followed by Canada, Chile, Brazil, Argentina, Mexico, Colombia, Venezuela, Peru and Ecuador, as illustrated in Table 3 (Economist Intelligence Unit and IBM Corporation, 2004:12).

The Latin American governments were reported to have also shown increasing support for e-commerce. For example, in Mexico, a law on digital signatures had just been passed. Most countries in the region were increasingly putting businesses online, and governments were creating policies and legislations that supported e-business development. Mobile penetration was also expected to grow, and the Internet was no longer the preserve of the rich (Economist Intelligence Unit and IBM Corporation, 2004:10). However, no country in Latin America featured in the world's top 25. The ranking placed most of the countries in this region in the 2nd tier, e.g. Chile (28th), Brazil (35th), Argentina (37th) and Mexico (39th). Within the Americas, the United States is a good example of the positive impact of an e-ready country, as reflected in its leadership in the e-commerce revolution. Consumer spending via e-commerce in the US leapt from Euro 1,104 million in 1997 to Euro 4,782 million in 1998; the European trends showed that Europe was still lagging far behind (Borg, 2004).

E-Readiness Status in Asia-Pacific

The Asian-Pacific region was ranked by the Economist Intelligence Unit in 2004 as the next best represented region in the e-readiness ranking after Western Europe. Singapore (7th) and Hong Kong (9th) were at the top, followed by Australia (12th), South Korea (14th), New Zealand (19th), Taiwan (20th) and Japan (25th). Most of these countries were characterised by access to broadband, telecommunications deregulation, and next-generation infrastructure development (Economist Intelligence Unit and IBM Corporation, 2004:14).

E-Readiness Status in the Middle East

In 2004, the Middle East's e-business development was only inching forward, especially when compared to Asia. Strict government control over Internet content and the over-regulation of service providers were reported to have been hampering development. Middle East governments had also, to some extent, restricted the entry of multinational companies into their markets. This was in a sharp contrast to Eastern Europe, where international competition was driving telecommunications development. Despite lagging behind in infrastructure and poor business environments, there were bright spots reported in 2004 by the

Economist Intelligence Unit, such as Israel (22nd), Turkey (45th), Saudi Arabia (48th) and Iran (57th).

E-Readiness Status in Africa

In Africa, e-readiness initiatives are being driven by the World Economic Forum and the e-Africa Commission's ICT Task Team which is responsible for developing NEPAD's ICT program and implementing its projects. The World Economic Forum and NEPAD's E-Readiness Policy Programme aim to help African countries develop e-readiness policies and to remove or reduce the policy obstacles that limit the use of ICT throughout the region. Some of the activities undertaken by the programme thus far include (Bridges.org, 2003):

- Establishing collaboration with the e-Africa Commission
- Identifying key actors and mechanisms to engage them
- Collecting information on basic e-readiness in African countries
- Creating a framework for examining challenges
- Proposing country groupings according to e- readiness levels.

Key findings with regard to studies undertaken by the World Economic Forum, NEPAD E-readiness Policy Programme and other agencies thus far, show that while the continent lags far behind its counterparts in the developed and transitional economies, there is growing unanimity among Africa's leaders and pan-African institutions on the benefits and impact ICT could have on a wide range of development issues. The ICT policy reform process is moving forward in almost all the countries in Africa, but there is no uniform level of progress across the continent. In general, the state of infrastructure development across the continent varies widely from country to country, with a considerable imbalance in the level of infrastructure roll-out across urban and rural Africa. Rural locations suffer at the expense of urban development. Africa also has an extensive and diverse range of ground level initiatives underway to promote and facilitate the use of ICT. Generally, there is evidence of growth in the e-commerce sector, most notably in South Africa. However for the rest of Africa, the cost of basic telephony and Internet connectivity remain disproportionately high across the continent, with the vast majority of Africa's population unable to pay for even basic access (InterAction, 2002).

The Information Society Index (ISI) measures have shown that Africa is very far behind other countries in the world as far as the ISI rating is concerned. ISI is an indirect indicator of a country's level of e-readiness. ISI provides a measure of the free and rapid flow of information across the world. It measures the information wealth of countries based on variables such as broadband households; mobile Internet users; software sophistication; wireless telephone subscribers; PC penetration; and education levels. In 2003, for instance, a ranking by the International Data Corporation involving 53 countries worldwide placed South Africa and Egypt (the only countries from Africa) at 30th and 47th positions respectively with regard to the index (Minton, 2003). Correspondingly, the 2004 e-readiness ranking by the Economist Intelligence Unit of sub-Saharan Africa rated South Africa as the most e-ready country at 32nd position out of 64 countries surveyed globally. Africa faces the challenges of low levels of adaptation of ICTs by administrators and businesses; resistance to change; and the poor status of existing infrastructure. In most countries in Africa, the networks most often used for data transportation are telephone networks. These are mainly digitised but badly adapted for high rate data transfer. The use of alternative technologies (with high bandwidth) such as the extended digital subscriber line (xDSL), local radio loops and optical fibre could be used to set up high rate data networks.

Despite the constraints facing Africa in its efforts to enhance e-readiness, most countries are engaged in the reform process of the telecommunications sector with encouraging results, especially with respect to the development of the cellular telephone system and the Internet. However, due to ineffective government policing of the liberalisation process, new telecommunications carriers have been prevented from competing against the de facto state monopolies. In addition, weak infrastructure development and high costs keep connectivity rates for voice and Internet services depressed (Economist Intelligence Unit and IBM Corporation, 2004:17).

IMPORTANCE OF E-READINESS ASSESSMENTS

Countries strive to attain some measure of e-readiness for various reasons. Foremost, they wish to become inclusive global information societies in which all persons are empowered to create, receive, share and utilise information for their economic, social, cultural and political development (Martin, 1997). By harnessing the potential of ICTs, national and local governments can now provide new and better responses to critical long standing issues such as poverty reduction, wealth creation, education, equity and social justice. Similarly, an understanding of the e-readiness of a country or community is essential for the provision of baseline information that can be used for planning and making comparisons across regions, countries, and organisations. E-readiness assessments can also be used as information gathering mechanisms that assist governments when planning strategies for ICT integration and also in making improvements on specific components of e-readiness (Rizk, 2004).

E-readiness assessments are also useful in that they can help governments understand and identify key and relevant ICT-based development opportunities. For example, in order to put ICT to effective use, a country must be 'e-ready' in terms of infrastructure; be able to provide access to ICT to the population at large; must have a legal and regulatory framework on ICT use; and benchmark progress and collaborations. Furthermore, an e-readiness assessment should lead to the development of a strategy or the preparation of an action plan that would address the opportunities and constraints identified in order to further the objectives of a country in the area of ICT. E-readiness assessments enable governments to measure and achieve realistic goals within an information society or information-based economy. It is important to develop and conduct e-readiness assessments so that the results can be leveraged to catalyze action, improve global competitiveness, and use limited resources wisely. Furthermore, a well-conceived e-readiness assessment can help map a country's regional and global position, improve competitive strengths, and promote those areas where it has an advantage over others. Understanding other countries' e-readiness strengths and weaknesses can also help a country leapfrog technologies and policy decisions so as to position itself ahead of its neighbours. E-readiness assessments can help stakeholders make difficult decisions on how to use scarce resources and how to turn existing strengths into new revenues. They can also reveal which bottlenecks are worthy of the investment of time and money to remove. Using a globally recognized e-readiness assessment method can be helpful in securing the necessary funding to develop an e-strategy and implement e-programmes (Docktor, 2002).

The value of e-readiness assessments is based on the notion that countries with pervasive information infrastructures that use ICT applications possess advantages for sustained economic growth and social development (The SADC E-readiness Task Force, 2002, p.12). E-readiness rankings provide companies wishing to invest in online operations with an overview of the world's most promising investment destinations. E-readiness rankings provide valuable insights into how governments can influence the rate and nature of the adoption of technology and applications (The Economist Intelligence Unit and IBM Corporation, 2004, p. 1).

CONCLUSION

This chapter has discussed the global status of e-readiness and provided a comparative picture of e-readiness across the world. North America and countries in Europe are leaders in e-readiness stature, and are significantly ahead of their counterparts in developing or transitional economies. The chapter highlighted the fact that countries with high levels of e-readiness are performing well in the digital economy. It also highlighted the value of e-readiness in enabling governments to measure their relative position against their counterparts both regionally and globally. E-readiness assessments are useful as information gathering mechanisms that assist governments when planning strategies for ICT integration and in making improvements on specific components of e-readiness. Such assessments help governments understand and identify relevant ICT-based development opportunities. The assessments also help businesses determine preferred investment destinations and more. The policy implications of the issues discussed in this chapter underscore the role of government in putting in place an institutional framework covering infrastructure and legislative mechanisms and regulations in order to create an environment conducive to improving e-readiness. How e-readiness is assessed, including the tools and methodologies used, are discussed in the next chapter.

REFERENCES

Borg, A. (2004, April 14-16). *Organisational and cultural transformation: The Malta experience*. Paper presented at the SADC E-Government Workshop, Gaborone, Botswana (pp. 4-37).

Bridges.org (2003). *Spanning the digital divide*. Retrieved on September 14, 2004, from http://www.bridges.org/

Commission of the European Communities. (2002, June 21-22). *E-Europe 2005: An information society for all*. An action plan presented at the Seville European Council, Lisbon (pp. 1-23).

Docktor, R. (2002). E-Readiness in 2002: *Defining and achieving your e-fitness goals*. Retrieved on August 12, 2004, from http://www.ip3.org/pub/docktor.htm

Economist Intelligence Unit and IBM Corporation (2004). *Scandinavia consolidates lead in fifth annual economist intelligence unit e-readiness rankings*. Retrieved on April 14, 2007, from http://www.store.eiu.com/index.asp?layout=pr_storyand-press_id=1260000726

InterAction. (2002). *World economic forum-NEPAD e-Africa commission e-readiness policy programme: Building capacity to narrow the digital divide in Africa from within*. Retrieved on August 12, 2004, from http://www.interaction.org/alpi/enews_NEPAD.html

Martin, W. J. (1997). *The global information society*. Aldershot: Aslib Gower.

Minton, S. (2003). Nordic nations still top information index. *The World Paper*. Retrieved on August 13, 2004, from http://www.worldpaper.com/2003/oct02/isi1.html

Rizk, N. (2004). *E-Readiness assessment of small and medium enterprises in Egypt: A micro study.* Cairo: American University. Retrieved on August 9, 2004, from http://www.sba.luc.edu/orgs/meea/volume6/Rizk.htm

SADC E-readiness Task Force. (2002). *SADC e-readiness review and strategy* (p. 1060). Johannesburg: SADC.

Chapter 7
E-Readiness Assessment Methods and Tools

MEASUREMENTS OF E-READINESS

E-readiness assessments are largely investigated at country-level across a number of sectors, and tend to adopt quantitative approaches that assign to countries' numerical scores depending on how well they have performed on specific components of e-readiness measures. A weighted average is calculated based on the relative importance accorded to these components in order to determine the level of e-readiness of countries (Rizk, 2004). The results of e-readiness rankings of countries are regularly published annually by some agencies. For example, the Economist Intelligence Unit (Economist Intelligence Unit, 2001) annually publishes a comprehensive list of countries on the basis of their measured e-readiness. The ranking categorises countries on the basis of their overall e-readiness, as calculated from 89 indicators

DOI: 10.4018/978-1-60566-420-0.ch007

across six weighted dimensions, namely connectivity, the business environment, consumer and business adoption, the legal and regulatory environment, supporting services, and social and cultural infrastructure. The result of the calculations is the classification of the world's largest economies on the basis of their perceived adopter category.

Globally, a number of countries have already attained significant levels of e-readiness, while others are at various stages of implementation. Countries that have attained significant levels of e-readiness are invariably situated in the developed world, and include among others, the United States, Canada, Singapore, Sweden, Japan, Finland, Britain, Norway and Australia. Most countries in the developing world are still striving to implement relevant infrastructures to attain levels of e-readiness sufficient enough to partake in the emerging global information economy.

MACRO E-READINESS ASSESSMENT TOOLS

There are several macro e-readiness assessment tools and methods that have been developed by various organisations. These organisations include, but are not limited to: Computer Systems Policy Project (CSPP); Centre for International Development, Harvard University, 2004; Economist Intelligence Unit, 2004; United Nations Development Programme, 2004; United Nations Conference on Trade and Development (UNCTAD), 2004; and SADC E-Readiness Assessment Task Force. Some of the most commonly used macro e-readiness tools include, among others: the Readiness Guide for Living in the Networked World developed by the Computer Systems Policy Project (CSPP); Network Readiness Index of the Harvard University's Centre for International Development (CID); E-readiness Rankings of the Economist Intelligence Unit; and the Technology Achievement Index of UNDP.

Each of these tools uses a different definition of e-readiness and methods for its measurement. Moreover, the e-readiness assessments are very diverse in their goals, strategies and results (Bridges.org, 2003). On average, however, the tools measure the level of infrastructure development; connectivity; Internet access; applications and services, or network speed; quality of network access; and ICT policy. The tools also measure:

- The ICT training programs in place
- Adequacy and availability of human resources
- Level of computer literacy
- Relevant content.

Largely, all the tools for e-readiness assessments have been designed for macro level assessments and are used to measure, for example, policy making, the state of Internet acceptance and growth, comparative analyses of countries, and stages of ICT development in countries (Dutta and Jain, 2004). In addition, the focus of each method or tool reflects the purpose for which it was designed. Moreover, the methods used in most macro e-readiness assessment studies vary from one tool to next. For example, APEC's Guide relies on questionnaire-based data, while Mosaic's methodology tends to be qualitative. Other organisations or agencies, such as CID, use a combination of questionnaires and raw data. McConnell's International provides a qualitative reference guide, whereas the World Information Technology and Service Alliance (WITSA) uses a survey-based method.

In selecting an appropriate tool for e-readiness, certain parameters need to be satisfied. The right tool should also depend on the users' goals, and requires users to choose a tool that measures what they are looking for and do it against a standard that fits their own view of an 'e-ready' society (Bridges.org, 2003). Accordingly, the tool should, among other things:

- Measure performance in all areas that are identified as important
- Be flexible enough to cover areas that it does not normally cover
- Provide assessments at both macro and micro level, i.e. from country down to the sectoral level
- Allow for the benchmarking of a community against other communities in each of the identified areas
- Be simple enough for a layman to understand.

Overall, although e-readiness methods and tools differ in definition, approach and objectives, they are all designed to gauge how ready a society or an economy is to benefit from IT and e-commerce.

THE NATURE OF E-READINESS ASSESSMENT TOOLS

E-Readiness Ranking Tool

The E-readiness Ranking Tool was developed by the Economist Intelligence Unit, which is the business arm of the Economist Group and publisher of the Economist magazine. Since 2000, the Economist Intelligence Unit regularly publishes an annual e-readiness ranking of the world's 60 largest economies. The E-readiness Ranking

Table 5. E-readiness ranking tool

Tool Name	Developer & countries studied	Intended application	Overall analysis
E-readiness Ranking Tool	The Economist Intelligence Unit and Pyramid Research, 2000 Has, since 2000, published annual e-readiness rankings of the world's 60 largest economies. In 2004, it expanded the number of nations covered to 64.	Intended to allow the ranking of governments to gauge the success of their technology initiatives against those of other countries. It also provides companies that wish to invest in online operations with an overview of the world's most promising investment locations.	The Economist Intelligence Unit is the business information arm of the Economist Group and publisher of the Economist magazine. Nearly 100 quantitative and qualitative criteria, organized into six distinct categories, feed into the e-readiness rankings. The criteria assess countries' technological infrastructure, their general business environment, the degree to which e-business is being adopted by consumers and companies, social and cultural conditions that influence Internet usage, and the availability of services to support e-business.

Tool defines a country's e-readiness as essentially a measure of its e-business environment, and bases its measure on a collection of factors that indicate how amenable a market is to Internet-based opportunities. The ranking allows governments to gauge the success of their technology initiatives against those of other countries. It also provides companies that wish to invest in online operations with an overview of the world's most promising investment locations (Economist Intelligence Unit and IBM Corporation, 2004: ii). The tool is described in Table 5.

E-readiness assessment using this tool makes use of nearly 100 quantitative and qualitative criteria organized into six distinct categories, namely:

- Connectivity and technological infrastructure
- Business environment
- Consumer and business adoption
- Legal and policy environment
- Social and cultural environment
- Supporting e-services.

Connectivity and technological infrastructure is given an overall weight value of 25% - it measures the access that individuals and businesses have to basic fixed and mobile telephony services, PCs and the Internet. Others are the affordability, quality and reliability of services.

Business environment is given an overall score of 20% - it evaluates the general business climate by screening 70 indicators covering elements such as the strengths of the economy, political stability, regulatory environment, taxation, competition policies, the labour market, quality of infrastructure, and openness to trade and investment.

Consumer and business adoption has an overall weight of 20% – it assesses how prevalent e-business practices are in each country, e.g. the share of retail commerce conducted online, and the extent of Internet use to overhaul traditional business processes.

Legal and policy environment is given an overall weight score of 15% - it assesses a country's overall legal framework and specific laws governing Internet use, that is the ease of registering new businesses, strength of protection of private property (intellectual property), and more.

Social and cultural environment has an overall weight score of 15% - it assesses literacy and basic education as preconditions to being able to navigate the web; a population's e-literacy - its experience using the Internet and its receptivity; and technical skills of the workforce.

Supporting e-services is given a weight score of 5% - it measures consulting and IT services and back office solutions, taking into account technological standards.

Readiness Guide for Living in the Networked World

This tool was developed by the Computer Systems Policy Project (CSPP), a public policy advocacy group comprising the chairperson and chief executive officers of the United States IT companies. It is a self-assessment tool that is designed to help individuals and communities determine how prepared they are to participate in the Internet (or networked) world. The tool measures the prevalence and integration of ICTs in homes, schools, businesses, healthcare facilities, and government offices, with an additional focus on competition between access providers, speed of access, and government policy. Measurements are divided into five categories, namely:

- Infrastructure
- Access
- Applications and services
- Economy
- Enablers - policy, privacy, security and ubiquity.

Table 6. Readiness for living in the networked world

Tool Name	Developer & countries studied	Intended application	Overall analysis
Readiness for Living in the Networked World	The Computer Systems Policy Project (CSPP), 1998 The tool is generally applicable; it does not examine any given country.	The CSSP self assessment tool is designed to help determine how prepared a community or a country is for participation in the networked world.	The CSSP is a public policy advocacy group and is comprised of chairpersons and chief executive officers of leading US information technology companies. The tool offers a series of 23 questions under 5 distinct groups: 1. The network (infrastructure) 2. Networked places (access) 3. Networked applications and services 4. Networked economy 5. Networked world enablers

Readiness for the Networked World

This tool, or guide as it is called, was developed by the Centre for the International Development (CID) at Harvard University, and was first published in 2000. It measures 19 different aspects of e-readiness, including:

- Availability of network access
- Speed and quality of network access
- Use of ICTs in schools, the workplace, economy, government and everyday life

Table 7. Readiness for the networked world

Tool Name	Developer & countries studied	Intended application	Overall analysis
Readiness for the Networked World	Centre for International Development (CID), Harvard University, 2001 Conducts analyses leading to the evaluation of a mix of developed and developing countries.	The tool is intended for government policy makers to assess the state of the networked readiness of a community. It is targeted at communities in developing countries seeking to define strategies to participate in the networked world.	This tool builds on the CSSP framework and has been developed in an academic setting. It presents a more balanced approach. The CID tool looks at 19 different categories or indicators that fall into 5 distinct groups: 1. Networked access 2. Networked learning 3. Networked society 4. Networked economy 5. Networked policy The tool analyzes each of the 19 indicators by classifying a country according to one of four categories or stages of development. The tool itself does not compare different communities, but rather presents an assessment based on the given inputs to various questions.

- ICT policy
- ICT training programs.

APEC's E-Commerce Readiness Guide

This tool was created by the Asian Pacific Economic Cooperation (APEC) Electronic Commerce Steering Group in 2000. Its aim is to help governments develop their own focused policies and adapt to their specific environment for the healthy development of e-commerce. The tool measures six categories of readiness for e-commerce, i.e.: basic infrastructure and technology (speed, pricing, access, market competition, industry standards, foreign investment, etc); access to network services (i.e. bandwidth, industry diversity, export controls and credit card regulation); and the use of the Internet (i.e. usage in business, government and homes). Other measures include:

- Promotion and facilitation (referring to industry led standards)
- Skills and human resources (i.e. ICT education and the workforce)
- Positioning in the digital economy (i.e. taxes and tariffs, industry self-regulation, government regulations and consumer trust).
- Basic infrastructure and technology
- Access to necessary services
- Current level and type of use of the Internet
- Promotion and facilitation activities
- Skills and human resources
- Positioning in the digital economy.

Table 8. APEC e-commerce readiness assessment guide

Tool Name	Developer & countries studied	Intended application	Overall analysis
APEC e-Commerce Readiness Assessment Guide	Asia - Pacific Electronic Cooperation (APEC), 2000 The tool has been used to evaluate various countries, such as Hong Kong and Malaysia	Helps member governments develop policies that promote e-commerce	The guide was created with the influence of industry to assist with the policy making efforts of governments. It is comprehensive in its coverage of e-readiness factors, with little analysis of the impact of past initiatives. It examines broader indicators of e-commerce, and these are encompassed in a series of questions that provide direction for the development of desirable e-commerce policies and the removal of barriers to electronic trade.

Risk E-Business Tool (Ready? Net.Go)

This tool was developed by McConnell International (2000) in collaboration with the World Information Technology and Services Alliance (WITSA), and was first released in August 2000. Its aim is to assess a national economy's e-readiness or capacity to participate in the global digital economy. It measures e-readiness in five areas:

- Connectivity (infrastructure and access and pricing)
- E-leadership (government policies and regulations)
- Information security (intellectual property, privacy, and electronic signatures)
- Human capital (ICT education and available skilled workforce)
- E-business climate (competition, political and financial stability, foreign investment and financial infrastructure).

Mosaic's Global Internet Diffusion Framework

This tool was developed by the Mosaic Group. Its aim is to measure and analyze the growth of the Internet throughout the world. It measures six dimensions: pervasiveness (per capita usage); geographic dispersion; sectoral absorption (use within major sectors of the economy); connectivity infrastructure; organizational infrastructure (state of the Internet service market); and sophistication of use. A country's movement along these dimensions is shaped by a set of determining factors, including metrics such as teledensity, PC density, per capita GDP, foreign and domestic investment, educational systems, and government policy. The analyses focus on major social, economic, political, legal and regulatory environments as they affect the countries' Internet status. The tool does not define an 'e-ready' society; rather it defines the readiness for growth of Internet infrastructure and its use. The tool uses a combination of statistics, narrative descriptions and comparisons to explain the growth of countries' Internet pervasion, focusing on the Internet statistics described above. Essentially, it is used to describe, in depth, the relative growth of the Internet in the given countries. The rest of the details about this tool are explained in Table 10.

This tool analyses internet diffusion along six indicators namely:

- Pervasiveness - level of use by individuals
- Sectoral absorption - level of use by academic, commercial, health and government sectors
- Connectivity infrastructure -quality and robustness of the underlying network

Table 9. Ready? Net.Go

Tool Name	Developer & countries studied	Intended application	Overall analysis
Ready?Net.Go	McConnell International, 2001 The 2001 report covered 53 countries with an emphasis on developing countries, although no geographic region was predominant.	Designed to assess a country's e-readiness or capacity to participate in the global digital economy. It aims to evaluate who is ready, i.e. which countries are enabling businesses, governments and citizens to flourish in the networked economy.	McConnell International is a consulting firm helping clients with technology policy and strategy formulation. Clients include governments, NGOs, multinational organizations, and private sector firms. It analyzes a country's e-readiness based on the following criteria: 1. Connectivity - whether networks are easy and affordable to access and use 2. E-leadership - the role that government and businesses play to promote the use of networked technologies in a country, and whether e-readiness is a national priority 3. Information security – whether the processing and storage of networked information can be trusted 4. Human capital – whether the right people are available to support e-business and to build a knowledge-based society 5. E-business climate - how easy is it to do e-business today?

infrastructure

- Organizational infrastructure - the number of and robustness of the organizations (for example ISPs) providing the infrastructure
- Geographic dispersion - how geographically dispersed the organizations providing the infrastructure are

Table 10. Framework of global diffusion of the Internet

Tool Name	Developer & countries studied	Intended application	Overall analysis
Mosaic's Global Internet Diffusion Framework	The Mosaic Group, 2001 The tool has been used in 25 countries in Asia, the Middle East, and Central Europe. It has also been used in Finland and Hong Kong.	Designed to assess the state of Internet diffusion in a country. It is useful for business stakeholders wishing to make use of and invest in the Internet; in debates about how to positively or negatively influence a policy's use and development; and to researchers studying the large-scale diffusion of complex interrelated technologies.	The tool is rooted in academia. The framework analyses the diffusion of the Internet in a country along six main lines

- Sophistication of use - how intense the adoption of the technology has been.

The tool addresses all stakeholders: individuals, businesses and government. It is particularly suited for the study of a given community. While the tool is useful for policy making, it makes the comparative analysis of nations' complex. The tool relies on a questionnaire as its primary source of data.

World Information Technology International Survey of E-Commerce

This tool was developed as a result of a survey carried out by the World Information Technology and Services Alliance (WITSA) in 2000. It defines a country's e-readiness according to whether the country encourages consumer trust in e-commerce security and privacy; offers better security technology, more trained workers and lower training costs; has less restrictive public policies and new business practices adapted to the information age; and offers lower costs for e-commerce technology. The tool functions as an international survey that solicits the views of information technology industry associations on the best ways to encourage the growth of electronic commerce. The tool focuses on the direct experiences of companies with e-commerce and their subjective views of what is needed to promote e-commerce. The questions cover a range of issues, including barriers to the technological industry, the role of consumer trust, problems with e-commerce technology, internal business practices that support e-commerce, workforce problems, taxes, public policy issues, and resistance from consumers (Bridges.org, 2001).

The assessment is carried out by asking technology companies a series of questions on their experiences with regard to barriers they come across when using e-commerce technology and recommendations for improving e-commerce. The output from the survey is usually presented in the form of a report that publishes and compiles answers from survey participants through charts and a narrative account of the answers. This tool has been used in 27 countries, with a good mix from both the developed and developing worlds (Dutta and Jain, 2004). Recommendations that have been made using this tool include working on solutions dealing with consumer trust, technology, creating a skilled workforce, public policy (such as trade, taxation, and encryption), new business models for ICT integration, and the cost of e-commerce. Further details about the tool are reflected in Table 11.

Table 11. International survey of e-commerce tool

Tool Name	Developer & countries studied	Intended application	Overall analysis
International Survey of e-Commerce tool	The World Information Technology and Service Alliance (WITSA), 2000 Has surveyed 27 states, with a good mix of developing and developed countries.	The tool seeks to determine: 1. What factors are most important for the deployment of e-business? 2. The degree of business and consumer use of e-commerce.	WITSA is a consortium of 38 IT industry associations that are often involved in policy-influencing activities. Their cross national perspectives are reflected in this tool; it has no special focus but highlights eight global issues in the development of e-business: 1. Trust, security and privacy 2. Technology 3. Workforce issues 4. Public policy 5. Taxation 6. Business processes 7. Costs 8. Consumer attitudes The tool is not meant to provide an understanding of country specific readiness measures of e-commerce. The results are important for policy makers seeking to understand the primary concerns around the development of e-business. Most of the factors considered in this tool are policy issues that fall under the remit of governments rather than the private sector.

SADC E-Readiness Assessment Tool

This tool was developed following the recommendations of the South African Development Community (SADC) Heads of States' declaration on ICT held in Blantyre, Malawi, from the 5th – 14th of August, 2001. This was followed by the setting up of an ICT Task Force which was charged with the responsibility of studying the e-readiness status of member states. The Task Force consisted of representatives from a few member states, academia and the private sector. The work of the Task Force was largely underpinned by the New Partnership for Africa's Development's (NEPAD) philosophy that new solutions befitting Africa need to come from the continent. The Task Force team formulated a framework for e-readiness assessment after conducting a review of other e-readiness studies. This framework was used to conduct e-readiness studies at macro level in all 14 SADC member countries.

The SADC E-readiness Framework measures e-readiness based on variables such as (SADC E-readiness Task Force, 2002): basic quality of telecommunications services and underlying infrastructure (e.g. the availability of landlines); legislation and policy frameworks; cultivation of ICT awareness and positive attitudes; general email and Internet use; education and training; PC penetration in businesses; affordability of telecommunications services; e-commerce adoption; content development; and the use of the Internet for global trade and other commercial activities. The framework also measures human factors that enable the use of technology. Such human factors include: education and training; access to media such as television and radio; relevant local content; the availability of electricity; ICT usage skills; and the application of different technologies. The tool has three levels of e-readiness which overlap as described below:

- **Fundamental level**: This level consists of the ground level enablers of e-readiness. It covers basic telecommunications, such as voice services, and basic underlying infrastructure and human factors that enable the use of technology. The latter include electricity, education, and access to media.
- **Middle level:** The second step in technology adoption involves access to more advanced telecommunications (data) services, including e-mail and the Internet, and the first stages of ICT infrastructure, with a specific focus on priority applications such as e-government, e-education, and e-health. The focus here therefore also rests on PC penetration in schools, clinics, community centres, government offices and businesses.
- **Advanced level:** The advanced level consists of latter stages and higher levels of ICT infrastructure and its applications, and implies the use of the Internet for global trade and other commercial activities. This calls for the existence or establishment of well developed banking infrastructure and other commercial and legal support mechanisms and policy frameworks.

More details about this tool are described in Table 12.

The World Bank E-Readiness Assessment Framework

The World Bank E-readiness tool provides a framework to support information and communication technologies in its client countries. The World Bank identifies six crucial areas wherein governments need to take action. These include (World Bank Group, 2002):

- Establishing general competition and legal frameworks
- Developing access infrastructure
- Developing e-commerce, content and convergence
- Promoting ICT application in government and the private sector
- Developing Human resources skills.

Table 12. SADC e-readiness framework

Tool Name	Developer & countries studied	Intended application	Overall analysis
SADC E-readiness Framework	SADC E-readiness Task Force, 2002 The framework was developed to conduct a study on 14 SADC member countries to establish their status of e-readiness.	The tool was developed to determine the level of e-readiness in SADC member countries according to competence in e-governance, e-services, e-business, ICT awareness, infrastructure, and policy and regulatory frameworks, among others.	The tool has three overlapping levels of e-readiness: fundamental, middle, and advanced.

Negotiating the Digital Divide Guide

This tool was developed by the Centre for International Development and Conflict Management (CIDCM) at the University of Maryland in partnership with the United States Agency for International Development. It was designed to assess the advancement of the Internet, with a particular focus on the sub-Saharan African region. Thus far, the tool has been used to assess the e-readiness of countries such as Ghana, Senegal, Kenya, China and Brazil (Dutta and Jain, 2004). The tool examines the deployment of the Internet as categorized into four stages:

Table 13. World Bank e-readiness assessment framework

Tool Name	Developer & countries studied	Intended application	Overall analysis
World Bank E-readiness Assessment Framework	The World Bank. The framework places an emphasis on the development of reliable, comprehensive information infrastructure in developed and developing countries.	1. The tool is based on the philosophy that countries with pervasive information infrastructures that use innovative information technology applications possess advantages of sustained economic growth and social development. 2. The tool focuses on the following ICT strategies: - Access infrastructure - E-commerce and convergence - Promoting IT and IT-enabled industries - E-governance - Human resources and capacity	The tool focuses on 6 key areas: 1. General competition and policy and legal frameworks 2. Access infrastructure 3. E-commerce, content and convergence legislation 4. Promoting ICT application in the government, private and social sectors 5. Industrial policies for the IT industry 6. Human resource skills

Table 14. Negotiating the digital divide

Tool Name	Developer & countries studied	Intended application	Overall analysis
Negotiating the Digital Divide	Centre for International Development and Conflict Management (CIDCM), University of Maryland.	The tool focuses on analyzing the development of the Internet in developing countries, particularly in African countries.	This tool is the work of CIDCM, an academic institution involved in conflict management and preventive diplomacy, in partnership with the U.S. Agency for International Development. The tool was designed not only to assess the advancement of the Internet, but also its enablement, with a particular focus on the sub-Saharan African region. CIDCM seeks a reusable model; however, the individual studies are nation-specific.

- Pre-commercial
- Commercial
- Competitive
- Consolidated.

The guidelines for the use of the tool recommend a combination of questionnaires and statistical analyses. It identifies the deployment of the Internet as being divided into four stages, namely:

1. Pre-commercial
2. Commercial
3. Competitive
4. Consolidated.

The tool also considers the ease and speed of negotiation between the different actors, i.e. individuals, businesses, governments and NGOs. Global Technology Index (GTI)

This tool was developed in 2002 by Dr. Howard Rubin, an IT consultant from Metricnet.com (a research firm that provides information consultations). The tool covers more than 50 countries spread across all the world's most important commercial zones. It has an even mix of developed and developing countries. The tool is meant to measure the economic dynamism, strength, and technological capabilities and potential of a country. This tool covers five factors to measure the GTI, namely (Dutta and Jain, 2004):

- Knowledge jobs
- Globalisation
- Economic dynamism and competition
- Transformation into the digital economy
- Technological innovation capacity.

More details about this tool are presented in Table 15.

International Records Management Trust (IRMT) E-Readiness Tool

IRMT is a UK-based non-profit organization that was established in 1989 to help build solutions for managing public sector records in developing countries. Over the last decade, it has played a pioneering role in addressing the relationship between records management and good governance. IRMT supports governments worldwide in building the infrastructure for managing public sector records. In partnership with public sector institutions, international donors, professional associations and academic institutions, it provides free records management training and resource materials, including the e-records readiness tool. The e-records readiness tool has been designed to enable governments to conduct high level assessments on key areas of e-records readiness in relation to other aspects of government in order to determine whether the records and information management infrastructure is capable of supporting government initiatives (IRMT, 2004).

IRMT's (2004) e-readiness tool uses a brief questionnaire to provide a risk assessment of e-records readiness in government at national and enterprise level. The tool consists of twelve components of e-records readiness. Areas addressed by the tool include: staff competencies in maintaining software and hardware; training programmes for information management staff; relevant ICT qualifications and experience; human resource strategies; ability of staff to recognize the importance of well managed information; availability of telecommunications infrastructure to support growing volumes of work; adequacy of electric power; formal records management practices; provision for the storage and retrieval of information; access to online/electronic information; information management policies & responsibilities; tools and procedures for information management; availability of information management products and technologies; internal and public awareness programmes

Table 15. Global technology index

Tool Name	Developer & countries studied	Intended application	Overall analysis
Global Technology Index	Dr. Howard Rubin, Metrinet.com, 2002.	Intended to measure the economic dynamism, strength, and the technological capabilities and potential of a country.	Metricnet.com is a data collection and distribution service and a division of the META group. The 2002 Global Technology Index is one of many such reports available at metricnet.com. The data for this report, however, is derived from statistics from independent data sources. The five factors used to measure GTI are: 1. Knowledge jobs 2. Globalization 3. Economic dynamism and competition 4. Transformation into the digital economy 5. Technological innovation capacity

of information management; compliance with information management procedures, i.e. security, backup, confidentiality, etc; guidelines for good practices in computer systems' security, backup and business continuity planning; adequacy of resources for e-records management; documentation standards and system engineering procedures for ICT; and ICT systems and formal requirements management processes.

The IRMT e-readiness assessment tool also provides guidelines for the management of electronic records; standards/formats for the storage and retrieval of data; information management policies on centralization/decentralisation; basic classification schemes for the information in place; policy on how information should be organized; ease of access to records or information; ICT strategies; presence of information management units; regular budgets to support information management; ICT awareness initiatives to new members of staff; published rules of access to information; and supportive legal and regulatory frameworks for information management. Other aspects of the IRMT tool include:

- Legal framework for e-commerce activities
- Freedom of information and protection of privacy
- Adequacy of a nation-wide ICT infrastructure
- Reliable electrical power sources.

E-Europe Model for Building Skills in Information Society

In a knowledge economy, knowledge and information are paramount; therefore citizens need new skills to adapt to rapidly changing life and work environments if they are to fully participate in society. In order to ensure that everyone effectively uses and benefits from ICT in life and work, communication and learning initiatives have come to be an important priority in Europe. The e-Europe Action Plan focuses on the following areas in terms of human resource development (e-Europe Action Plan, 2004): digital literacy; access and effective use of ICT; addressing digital divide issues; ICT skills development; lifelong learning, particularly for those whose access to these technologies is limited because of distance, social situation or special needs; and information literacy to cope with the vast amount of information being generated.

World Economic Forum Framework

The World Economic Forum Framework is summarised in Table 16 below.

MICRO E-READINESS ASSESSMENT TOOLS

While there are several tools available for measuring e-readiness at macro or national level, there is a paucity of tools for sectoral assessments. There is a fallacious assumption about ICT and small-sized enterprises that the significance of ICT at national level is the same as the significance at an individual firm level. However, e-readiness at sectoral level is influenced by a different set of economic, social and geographical circumstances; therefore similar strategies cannot be applied wholly to such an environment without being tailor-made to suit its specific needs for the short, medium and long term. Industry sector differences and management characteristics and the supplier-customer environment of firms are some of the factors that influence the degree of e-readiness of SMEs.

A few studies globally have attempted to evaluate e-readiness from a micro-perspective. Only a small number of e-readiness assessments at micro-level with respect to ICT adoption in SMEs have been undertaken in the United States, Australia, Europe and Asia. Micro e-readiness studies are important as they capture many of the factors that may escape macro-analyses, and hence offer a more accurate picture of the phenomena. Rizk (2004), in a study on the Egyptian SMEs in the textile industry, assessed the firms' levels of connectivity, awareness and use of ICTs in marketing, production and management. Rizk ranked the level of e-readiness of each firm by averaging all the variables under study according to low, medium and high scores. Using the same categorisation, Rizk was able to show the level of e-readiness of each firm on a given variable. Rizk also assessed the effect of a firms' size on their preparedness to embrace ICTs and identified the nature of and

Table 16. World Economic Forum Framework

Tool Name	Developer & countries studied	Intended application	Overall analysis
World Economic Forum Framework	World Economic Forum.	The tool emphasizes the importance of telecommunications deregulation and liberalization and the benefits it can bring to the ICT industry.	The tool focuses on three main areas: 1. Regulatory environment –macro-economic, industrial sector specific policies. 2. Infrastructure – computer hardware, satellite connections, etc. 3. Ground level initiatives such as public access to ICT by communities, and entrepreneurship.

the barriers to the ICT use of these firms. Rizk's Micro E-readiness Assessment tool is shown in Table 17.

Agriculture and Food Canada (2001), in a study on the e-readiness of SMEs in the agricultural sector in Canada, defined the degree of e-readiness of a business along a continuum on a scale of 1 to 7, as follows:

1. Limited use of computers
2. Reasonable use of computers
3. E-mail and access to Internet
4. Corporate website (info only)
5. Interactive corporate website
6. Taking orders/ordering over the Internet
7. Taking payments on the Internet.

The logic behind the seven point scale is that it can allow for a grouping of SMEs along a continuum of current and expected degrees of 'e-readiness'. Under this scale, companies that undertake activities falling under levels 6 and 7 can be said to have achieved a high sophistication of e-readiness. Moreover, it is assumed that the degree

Table 17. Rizk's micro e-readiness Assessment tool

Tool Name	Developer & countries studied	Intended application	Overall analysis
Rizk's Micro E-readiness Assessment tool	Nagla Rizk, 2002 The tool has been used to assess the e-readiness of micro-enterprises in Egypt's textile sector.	An attempt to model e-readiness at micro level by adapting and modifying macro e-readiness tools.	The tool covered 6 areas, namely: 1. E-infrastructure 2. Human capital 3. Connectivity 4. ICT awareness 5. ICT usage 6. Barriers to ICT usage

of e-readiness is a function of a firm's, CEO's and stakeholder's characteristics; and the nature of stakeholder interaction. Other measures include:

- Supplier environment
- Customer environment
- Present day e-business and e-readiness of the company
- Perceived costs and benefits
- Barriers for adopting e-commerce.

A summary of the Agriculture and Food Canada E-readiness Assessment Tool is reflected in Table 18 below.Other micro e-readiness assessment tools are presented in Tables 19 and 20.

STRENGTHS AND WEAKNESSES OF E-READINESS ASSESSMENT TOOLS

Each of the existing e-readiness tools has its own strengths and weaknesses. For instance, APEC's e-commerce readiness guide's strengths lie in its comprehensiveness and ease of use. However, its major weakness, as with all the other assessment tools, is that it is meant for macro level assessment and is not suitable for the assessment of sectoral e-readiness (for example of SMEs). The tool also does not give enough attention to information access - a critical element for SMEs wishing to partake in the digital economy.

From an information science point of view, the way information is structured,

Table 18. Agriculture and food Canada e-readiness assessment tool

Tool Name	Developer & countries studied	Intended application	Overall analysis
Agriculture and Food Canada Micro – E-readiness Assessment Tool	Agriculture and Food Canada, 2001 The tool has been used to study the e-readiness of SMEs in the food and agricultural sector in Canada.	Intended to measure e-readiness on a 7 point scale that allows the measure of not only the current degree of e-readiness, but the expected degree as well.	The tool defines the degree of e-readiness of businesses along the following continuum of seven levels **Level:** Limited use of computers **Level 2:** Reasonable use of computers (no e-mail) **Level 3:** E-mail and access to the Internet **Level 4:** Corporate websites (info only) **Level 5:** Interactive corporate websites **Level 6:** Taking orders/ordering over the Internet **Level 7:** Taking payments/ paying over the Internet

Table 19. SME survey tool (South Africa)

Tool Name	Developer & countries studied	Intended application	Overall analysis
SME survey tool (south Africa)	Hewlett Packard, ABSA bank and Price-WaterhouseCoopers in South Africa.	Used to study the annual e-readiness of SMEs in South Africa.	The tool measures the competitiveness of various firms in the following areas: 1. Productivity 2. Profitability 3. Cost reduction 4. Market share The tool measures: 1. Use of ICTs 2. ICTs implemented 3. Support provided 4. Barriers to ICT use 5. Services which ICTs are used for 6. Impact of ICTs

Table 20. Holistic approach (Ireland) tool

Tool Name	Developer & countries studied	Intended application	Overall analysis
Holistic approach (Ireland)	Ramsay et. al (2003)	Designed to measure the e-opportunities of service sector SMEs in Ireland.	The tool is divided into three stages: **Stage 1:** Covers organizational demographics such as: 1. Types of services offered 2. Market environment 3. ICTs, i.e. attitudes to e-business or awareness of benefits, actual Internet access, and Internet usage. **Stage 2:** Covers web sophistication functionalities such as: 1. Nature of services 2. E-commerce usage 3. Level of sophistication and functionality of websites **Stage 3:** Covers case studies including: 1. Internal e-enablers 2. Internet commerce usage 3. The real barriers

stored, retrieved and manipulated through the formalism of information modelling is of critical importance in the e-readiness assessment tools. It is important that the assessments of information readiness (as part of e-readiness) within SMEs consider how the information systems available connect the users to the information sources relevant to their needs, with expectations that the users would be able to retrieve and internalise that portion of information required to achieve knowledge and satisfy their needs. However, most existing macro e-readiness tools have relegated the importance of information to the periphery and instead put more premium on infrastructure, legal and policy frameworks, the business environment and human resources.

The existing tools apply mostly quantitative as opposed to qualitative methods. Qualitative tools would focus more on human factors, such as perception, attitudes, acceptance of ICT, and the like. The e-readiness tools available also do not assess the impact of e-readiness on enterprises or organisations. In general, the existing methods and tools were developed to measure the e-readiness of countries and make comparisons based on the average scores of various variables. The tools seem to operate on the principle of 'one-size-fits-all'. The existing tools also fall short of integrating the various metrics of existing e-readiness assessment models. For example, whereas Harvard University's model looks at how ICTs are currently used in society, APEC's tool focuses on government policies for e-commerce (Bridges.org, 2001). APEC's tool also provides very little coverage of technology in society; locally relevant content and products; and the training required for people to perform electronic business transactions. Moreover, APEC's survey measures specific environments against a single standard of how a country should be in order to promote e-commerce, implying that the same structure of e-commerce is applicable to any country.

The rest of the e-readiness tools seek to measure e-economy metrics and e-society indicators. The e-economy assessment tools look at the ability of ICT to impact on the economy, while e-society assessment tools look at the potential impact of ICT on greater society (Grigorovici et al., 2003:26). In general, a common parameter in macro e-readiness assessments is the inclusion of some measure of physical infrastructure/usage (e-infrastructure) and education (including knowledge of ICTs). These represent the lowest common denominator for assessing the macro e-readiness of countries, and are complemented by policy and economic environment settings.

The macro e-readiness assessment tools have also been criticized for ignoring the internal divides existing between developed and developing countries. For example, the Harvard framework of e-readiness proves insufficient because it limits itself to gathering certain facts about the penetration and use of ICT (mostly in the form of global averages), while completely neglecting socio-economic contexts and historical and cultural backgrounds. The Harvard framework does not take into account the more general effects of an illiterate society. It is also biased in favour of the middle class consumer of developed countries, and does not identify or offer bridging solutions that could benefit more people, especially the illiterate and the poor. Harvard's e-readiness assessment assumes that one is either connected to the Net or not, an assumption that does not apply to large portions of developing countries. The Harvard method also ignores the fact that the Internet boom in developed countries has had a solid background sustained by the consistent development of in-house, applied ICT, accompanied by the systematic re-engineering of business processes in both the public and private sectors, even long before the public form of the Internet appeared (Arce and Hopmann, 2002:1, 22).

The macro e-readiness tools are limited in scope and also lack detailed descriptions of how to use the tools in practice. Across the board, most tools simply state what factors are necessary for e-readiness, but do not provide any detailed rationale as to why (Bridges.org, 2001).

CHOOSING THE RIGHT E-READINESS ASSESSMENT TOOLS

Choosing the right e-readiness assessment tool depends on the users' goals and needs. For example, if the goal is to assess a country's preparedness for e-commerce, APEC might be the best-suited tool, first because it is quite detailed with its 100 questions, and second because it is straightforward to use. However, if the goal is to assess the current level of technology in a region in order to forecast future technological growth, there are a number of options. Questionnaires, such as the CID and CSPP guides, measure a wider set of statistics, such as bandwidth, reliability, pricing, and usage in homes, schools, businesses, and government. However, these are very rough measurements because the granularity is low for answers – users can only choose from one of four multiple choice answers.

CSPP and CID are therefore considered best for a quick but rough assessment of the use of technology. On the other hand, if the goal is to understand the relative roles of political, economic, and social factors in technology's growth and use, then the use of statistical methods, such as McConnell's International's method, are better suited. If the goal is to understand why particular countries progress differently, the case-study method used by Mosaic and CIDCM may be appropriate. However, if the goal is to assess the effects of the technology on the lives of people and consider how widely the technology is being used, then the digital divide reports may be more helpful. The digital divide reports focus on education, local and relevant content, and the effective use of technology across society (Bridges.org, 2001). Clearly, none of the existing e-readiness assessment tools is exhaustive with regard to the parameters one would wish to measure in any e-readiness assessment study. Bridges.org (2001) noted that it might be faulty to use a single standard of measurement at all. This is because there has been no single social, political, and/or economic model identified as the most successful in harnessing information technology. The attempt in this book to model an integrated tool (described elsewhere in this book) could help partly fill this lacuna.

CONCLUSION

This chapter discussed various e-readiness assessment tools. Largely, the existing e-readiness assessment tools are designed to investigate macro levels and not micro sectoral entities such as SMEs. Macro e-readiness assessment tools are diverse, as reflected in the various definitions and methods of assessment. The choice of the tool therefore mainly depends on its purpose, and who is carrying out the investigation. The existing tools focus more on ICTs and the business environment, but pay little attention to information access, which is an important variable in empowering small-sized enterprises in the global competitive networked world. The macro e-readiness tools that exist fail to address the impact of e-readiness on business organisations, and often do not provide suggestions on what is needed to move a country, for example, up the e-readiness ladder. The existing macro e-readiness assessment tools are also largely quantitative, with their methods of data collection mainly based on short survey questionnaires, and as such, very little information is captured on people-centric issues such as perceptions, attitudes, skills, and awareness. Without these detailed qualitative aspects of e-readiness, a void exists in decision making processes on how improvements can be made to enhance the status of business organisations. Furthermore, the existing e-readiness tools were designed to model situations in developed countries, and therefore their application to different environments (for example, to developing countries) may not reflect accurate results.

This chapter has significant implications for governments and managers of small-sized enterprises. Governments must deploy suitable e-readiness tools to ensure that accurate results are obtained to assist in planning and policy formulation. Without accuracy, resources may be wasted by investing in areas that are unprofitable. Similarly, managers of SMEs must be conversant with appropriate e-readiness tools if they are to effectively use e-readiness outcomes in order to identify their actual competitive position in the world.

REFERENCES

Agriculture and Food Canada. (2001). The e-readiness of Canadian Agriculture and Food Canada SMEs. Retrieved on March 21, 2004, from http://www.agr.gc.ca/misb/fb/food/ecommerce/eready/introduction_e.html

APEC. (1995). *APEC survey of technology information needs and training opportunities for SMEs, APEC Committee on SME*. Manila, Republic of the Philippines: Coordinating Agency, Dept. of Trade and Industry. Retrieved on August 21, 2004, from http://www.actetsme.org/index.html

Arce, M. E., & Hopmann, C. (2002). *The concept of e-readiness and its application in developing countries: Methodological problems and results for the e-readiness of Nicaragua* (pp. 1-25). Managua: Nicaragua National Chamber of Industries.

Bridges.org. (2001). *Comparison of e-readiness assessment models: Final draft.* Retrieved on July 16, 2003, from http://www.bridges.org/eredainess/tools.html

Dutta, S., & Jain, A. (2004). *The networked readiness of nations: Global information technology report 2003-2004.* Retrieved on August 25, 2004, from http://unpan1.un.org/intradoc/groups/public/documents/APCITY/UNPAN014654.pdf

Economist Intelligence Unit & IBM Corporation. (2004). *Scandinavia consolidates lead in fifth annual economist intelligence unit e-readiness rankings.* Retrieved on April 14, 2007, from http://www.store.eiu.com/index.asp?layout=pr_story-andpress_id=1260000726 e-Europe Action Plan. (2004). *Building skills for the information society.* Retrieved on November 8, 2004, from http://europa.eu.int/information_society/edutra/skills/index_en.htm

Economist Intelligence Unit Limited. (2001). *E-readiness rankings.* New York: Economist Intelligence Unit.

Grigorovici, D. M., Schement, J. R., & Taylor, R. D. (2003). *Weighing the intangible: Towards a framework for information society indices.* University Park, PA: Institute for Information Policy, Pennsylvania State University.

IRMT-International Records Management Trust. (2004). *The e-records readiness tool.* London: IRMT.

McConnell International. (2000). Risk e-business: Seizing the opportunity of global e-readiness. Retrieved on September 14, 2004, from http://www.mcconnellinternational.com/ereadiness/EreadinessReport.htm

Rizk, N. (2004). *E-readiness assessment of small and medium enterprises in Egypt: A micro study.* Cairo: American University. Retrieved on August 9, 2004, from http://www.sba.luc.edu/orgs/meea/volume6/Rizk.htm

SADC E-readiness Task Force. (2002). *SADC e-readiness review and strategy* (p. 1060). Johannesburg: SADC.

UNCTAD. (1998). *Handbook on foreign direct investment by small and medium-sized enterprises: Lesson from Asia.* Geneva: UNCTAD.

World Bank Group. (2002). *Building Africa's communication infrastructure* (p. 105). Washington, D.C.: The World Bank.

World Information Technology and Services Alliance. WITSA. (2000). *The world information technology and services alliance international survey of electronic commerce.* Retrieved on April 12, 2007, from http://www.witsa.org

Chapter 8
E-Readiness of SMEs

INTRODUCTION

Increasingly, SMEs are achieving and sustaining competitive advantages using ICT that are propelling and accelerating the globalization of businesses. The growth of the Internet has created a global, cost-effective platform for businesses to communicate and conduct commerce. The Internet is making it possible for SMEs to enjoy the benefits that were once only afforded to larger businesses. A study by Ramayah et al. (2003) in Malaysia revealed that most SMEs in that country are increasingly embracing e-commerce and spending increasing amounts of money on information technology, with subsequent higher revenues. SMEs in the digital economy have significant advantages over larger competitors (Dejonckheere et al., 2003) arising out of:

DOI: 10.4018/978-1-60566-420-0.ch008

- Increased outsourcing of information activities, resulting in new business opportunities
- The low degree of capital intensity of many e-businesses, which has resulted in relatively low start-up and exploitation costs
- Enhanced speed of decision-making and innovativeness
- Electronic networking and clustering, which allow SMEs to combine the advantages of being small-sized with the various benefits of large scale enterprises.

Despite the increasing awareness amongst SMEs of the importance of ICT to enhance their businesses, most are only able to implement basic online transactions such as daily sales, customer liaison, advertising company products, soliciting prices and communicating via e-mail. In developing countries, it would seem SMEs are not under pressure to move most of their transactions online, in part because most clients, including governments, do not have the requisite infrastructure. Selhofer (2003) observes that e-business requires appropriate network access (including sufficient bandwidth), internal hardware and software solutions, and procedural and managerial readiness to deal with online transactions, from simple web presence to the fulfilment of customer orders. The various levels of transacting business online by SMEs can be grouped into four progressive stages as follows:

Stage 1: Websites at this stage offer goods and services in the form of static information on web pages. Examples of the kind of information one finds here include online brochures and information services. At this point, there is usually no individual ordering or other online services.

Stage 2: Websites are interactive, with dynamic feedback and individualised product information. Here one comes across real time statistics, database search options, product catalogues and e-mail interfaces.

Stage 3: The websites in stage 3 provide interactive features to customers, with the possibility of online ordering and payments with a credit card. Typical examples of applications and information at this stage are online shops; online markets (shopping malls); airline or theatre ticket sales; online contracts, etc.

Stage 4: Websites provide more sophisticated interactive features. Examples are suppliers of music on demand and software distributors, and general online sales, transactions and distribution.

Most SMEs in developing economies are reported to be at stage 1, with a few at stage 2, suggesting that their e-readiness status is very far from allowing them to effectively participate in the global business environment.

E-READINESS DECISION ADOPTION PROCESSES AMONG SMEs

There are various factors that influence the e-readiness adoption decisions of SMEs. One such aspect is the degree and level of communication that is required across the enterprise. It is generally expected that firms with dispersed, multi-location facilities would be more advanced in their use of the Internet as a means of communication. Firms in a single location do not require any sophisticated method to communicate with their employees; hence single location firms, in relation to multiple location firms, generally rank lower in terms of their degree of e-readiness. The type of a firm's ownership may also be linked to the degree of the firm's e-readiness. The degree of a firm's e-readiness tends to be highest if the firm is publicly owned, followed by firms that are private but non-family owned, and lastly by firms that are family owned (Agriculture and Food Canada, 2001).

The characteristics of senior management, especially of the chief executive, might also have an important impact on the level of e-readiness. E-readiness assessments of SMEs in the US and Canada have shown that the lower the education level and higher the age and tenure of the chief executive, the lower is his/her interest in enhancing a firm's degree of e-readiness. Comparatively, a chief executive with a higher level of education tends to promote a higher degree of e-readiness in his/her firm. Generally, small-sized enterprises rank lower in terms of their degree of e-readiness compared to larger firms. A firm's decision to undertake an e-readiness investment is often first based on the amount of investment required to become e-ready, and second on the perceived benefits that can be gained by doing so (Agriculture and Food Canada, 2001). The e-readiness adoption decision is also influenced by the nature of the supplier, customer interactions and the environment. In particular, the number of suppliers, their geographical proximity, and the number and complexity of transactions and contractual relationships influence e-readiness adoption decisions.

The simpler and smaller the number of transactions, the lower the business's reasons for adopting a very high degree of e-readiness. Firms will have a higher propensity for e-readiness if they are dealing with suppliers that are heavy users of technology and who encourage them to use the Internet. Furthermore, the farther away the major suppliers, the greater the logic behind using electronic means to transact with a firm's customers; consequently, the greater the influence on the degree of the firm's e-readiness. The adoption of the Internet by SMEs can also be influenced by the e-readiness status of the customers they transact business with. Thus, the impetus to improve the degree of e-readiness can stem from the SMEs side of the equation and/or from the customer side.

The degree of e-readiness has been found to vary according to a firm's size as measured by annual sales. In Canada's Agricultural sector, firms with annual sales greater than US$ 75 million have the highest percentage of respondents with high scores in their degree of e-readiness, while firms with annual sales of less than US$ 1 million have the highest percentage of respondents ranked with low levels of e-readiness. The degree of e-readiness also varies by size as measured by the number of employees. Firms with less than 50 employees tend to have the highest percentage of respondents with low scores in their degree of e-readiness, while firms with more than 50 employees have a higher percentage of highly ranked respondents in terms of the same (Agriculture and Food Canada, 2001).

The growing impetus for the adoption of ICTs in SMEs around the world is also being driven by the falling costs of hardware and software, more user friendly software, easy to install personal computer networks, and a number of other related factors. The most widespread use of the new technologies exists largely in enterprises that have adopted the most thorough range of work organisation changes, such as the decentralisation of decision making and the organisation of work into semi-autonomous task-oriented teams. E-readiness adoption is affected by a multiplicity of other factors. The World Economic Forum (2002:15-18) suggested that three major factors motivated developing countries' decision-makers to improve e-readiness and promote the adoption of ICT in their countries. The first was that ICT promises enormous benefits as part of the solution to economic and social problems by empowering people in developing countries and enabling them to better themselves. Another motivation is that countries face the threat of being left even further behind if they do not address the growing digital divide, both between and within countries. There is also growing pressure from the international community, especially foreign donors and lending agencies, for recipient countries to integrate ICT into their development aid programmes.

As already mentioned, e-readiness adoption decisions are also influenced by the traditional costs and benefits of economic investment, the complexity of technology, and the e-readiness of the SMEs' supplier-customer chain. Southwood (2004a) observes that in Kenya, for example, the country's national carrier (Kenya Airways), driven by its vision of becoming a world class airline, realized after market analysis that it suffered various handicaps by having its market base in Africa, key among which was the lack of access to proper ICT infrastructure. The company acknowledged that it had a customer base outside Kenya who perceived buying tickets over the Internet as normal. Recognizing that the travel market was moving into cyberspace, the airline came to realize that it had to move with its competitors into that environment. The airline was also driven by lower Internet tariffs to start online ticket sales.

INTERNATIONAL E-READINESS STATUS OF SMEs

Levels of e-readiness have implications for the capacity of SMEs to partake in the global digital economy. Generally, most governments across the world have been investing in ICTs to enable SMEs to enhance their productivity and competitiveness. Information Society Index measures have shown that half of the countries from Western Europe, North America and Australasia, dominate other countries in e-readiness leadership. There was a clear correlation between information wealth and economic prosperity. The Information Society Index is a new measure developed to understand the free and rapid flow of information across the world (Minton, 2003).

E-Readiness of SMEs in North America

Countries in North America with well developed ICT infrastructure invariably have most of their SMEs increasingly going online. For example, a survey of e-readiness of SMEs in the agricultural sector of the United States showed that a significant number of firms were poised to reach their highest level of e-readiness in the near future. Similarly, Canada was, in 2003, the world's most connected country and a global leader in the digital economy, with approximately 85% of its SMEs and larger enterprises using the Internet. Findings on the degree of e-readiness of the United States' SMEs within the agricultural sector revealed that 69% of the SMEs had websites, and 31% accepted credit cards as payment for business to business (B2B) transactions (Consulting and Audit Canada, 2004:12).

In 1998, Canada had 70% of its medium-sized businesses connected to the Internet and with a website compared to 76% in 1999. Similarly in 1998, only 6% of its small-sized enterprises had a website, compared to 11% in 1999. The percentage of Canada's SMEs that were connected to the Internet rose from 15% in 1996 to 69% by 2000. Increasingly, SMEs in Canada have become adept at using the Internet for communication purposes. The study on the ICT use amongst SMEs in the agricultural sector in Canada revealed that a larger majority of the firms were using e-mail as a communication medium, and approximately 25% of the firms had a website. In addition, 6% of the firms processed orders or accepted payments over the Internet. The results of the study further revealed that 60% of the companies enabled online purchases. It was expected that over the next two years, 85% of the firms would increase their online transactions (Agriculture and Food Canada, 2001).

Canada has taken deliberate efforts to enhance the participation of its SMEs in the global digital economy. Canada's strategies business site (http://strategies.ic.gc.ca/) is maintained by Industry Canada as a one-stop kiosk for Canadian firms seeking to find information about carrying out business in Canada and abroad. It includes important information for foreign firms wishing to invest in Canada. Various databases are found on this site, including a business registry. Through another portal - the BusinessGateway.ca - business start-ups can carry out an online entrepreneurial self assessments, identify niche market opportunities, select appropriate technology solutions, prepare a business plan, find skilled staff, identify financing, and incorporate a company (Consulting and Audit Canada, 2004).

E-Readiness of SMEs in Europe

Within Europe, the e-Europe Action Plan was endorsed by the member states of the European Commission in June 2000. The plan outlined the measures that the member countries were to take in order to achieve an information society. Among the measures of the plan were encouraging SMEs to go digital through coordinated network activities for the exchange of knowledge on best practices; e-commerce readiness; and mainstreaming e-commerce into SMEs' business strategies. In addition, the Action Plan envisaged that by 2005, Europe should have a dynamic business environment with a public policy to provide added value in the area of legislation affecting e-business (Commission of the European Communities, 2002:8). By the end of 2003, the European Union member states had established a European e-business support network, federating existing European national and regional players in this field with a view to strengthening and coordinating action in support of SMEs in the field of e-business. The commission was further expected to foster geographical and sectoral clusters of SMEs working online to encourage innovation in e-business, the sharing of best practices, and the promotion of guidelines and standards. Furthermore, the commission developed a comprehensive policy in the field of e-business aimed at establishing the Internet market for the information society, and launched a 'Go Digital' initiative to help SMEs better use e-business.

A number of countries within Western Europe have since created a thriving and elaborate SME sector using ICTs, a good example of which is Turkey. The Turkish government's SME efforts are directed at forming platforms that prepare the SMEs for e-finance. Efforts are being made to create awareness of the benefits of access to the right information at the right time via the Internet. Third parties are creating electronic market platforms for many products, from raw materials to semi-finished goods, bringing together partners, suppliers, purchasers and intermediaries. They are also allowing electronic communication and coordination across the supply chain (Gungen, 2003). Turkish banks have set up a strategy focusing on the specialised needs of SMEs. Services such as a credit scoring system to improve risk management, credit cards for the settlement of B2B transactions, Internet credit card processing services, electronic bill payment systems, and visa authentication payment programmes, are provided to SMEs. Turkey's SME sector is actually primarily serviced by Disbank, a private bank that caters for its ICT needs. The services that it provides include software and hardware for accounting, bill production, e-banking and insurance. The bank also offers PC training, technical support, and free advertising on the bank's website.

The government of Turkey is further involved in providing trade related information to SMEs aimed at increasing their ability to do business electronically and help them develop international trade. For instance, the Trade Points Programme, a Turkish arm of UNCTAD, facilitates access to the latest ICTs by making them available to trade operators in developing countries and to SMEs worldwide at reasonable cost. The National Productivity Centre also provides consultancy services to SMEs at very low costs. The ISP and ICT companies have special offers for SMEs, such as special payment options, leased and tailored technology training, unlimited Internet access, Internet credit card processing, and the management of credit card transactions. A typical ISP offer would tend to include unlimited Internet access, a static IP address, domain name, web space and e-mail address. There are also special portals for SMEs through which financial news and interpretations, macroeconomic data and information can easily be accessed (Gungen, 2003). For example, Koc Group and Microsoft Turkey have jointly established a business portal known as kobiline.com that targets SMEs. The site provides expert advice, consultations, a product catalogue, company information, order taking, product search, seller search, free email, and web space. Further on, there is toptanpazaryeri.com, also a business portal targeting SMEs and a major B2B and contact centre of Turkish international trade.

In Malta, the government has made a commitment to nurture existing and new IT businesses, whether they are locally or overseas-owned. A special package of incentives, drawn up with the specific intent of building up this sector as a supporting pillar for the emergent information economy, has been put in place (Borg, 2004:12). The government has also undertaken to ensure that any barriers and unnecessary regulations that prevent business growth are removed, and that the necessary aids for investment in this field are readily available. Malta's government is also involved in implementing a nationwide capacity building exercise that aims to achieve widespread digital literacy on a national scale via targeted initiatives, placing particular emphasis on the ability to use the Internet.

The government of Estonia in the former Soviet Union Republic has also been promoting connectivity and building ICT infrastructure in various sectors of its economy to enhance development. A report of e-readiness on 42 countries rated Estonia among the top 20 most advanced nations. The government of Estonia's e-readiness and IT related expenditure over the 7 years prior to 2000 was estimated at 1% of the total national budget. The government was also involved in reforming the institutional infrastructure in the SME sector, the objective being to increase efficiency by streamlining the various functions and responsibilities (Trend Chart on Innovation, 2000).

E-Readiness of SMEs in the Middle East

The Middle East has one of the most thriving, emerging SME sectors in the world. As an example, a study that was carried out in Jordan between December 2002 and February 2003 to determine the level of technology and Internet adoption amongst SMEs in the country, found that 78% of the country's SMEs used computers, 67% used the Internet, and 52% had LANs set up in offices. The study revealed that SMEs in Jordan were supported by the National Fund for Enterprise Support, whose aim it is to help SMEs become more efficient and capable of competing regionally and globally. The support for SMEs was facilitated through a financial management development project subsidy. Overall, Jordan's SMEs across several different economic sectors showcased satisfactory adoption of ICTs (PRWeb, 2003).

E-Readiness of SMEs in Asia-Pacific

In Asia, a study on the ICT infrastructure in the Philippines, focusing on the country's e-readiness at both macro and micro level and with a special interest in the SME sector, established that strategic areas key to technological development in the country include ICT infrastructure and its systems, the development of an e-business environment, human resources, IT education, and legal and policy frameworks (Press Release, 2002). The government of the Philippines enacted an e-commerce Act in 2000 and became one of the first countries globally to put in place a law governing e-commerce transactions (Sachs, 2003).

In the case of Japan, the Information for Development Program (InfoDev), with support from government, launched a three year incubator initiative in 2002 aimed at enhancing the entrepreneurship, innovation and competitiveness of small-sized enterprises in developing countries. The initiative meant to strengthen the incubators of small-sized enterprises and similar programs in recipient countries, and through them, stimulate the emergence and growth of innovative, highly competitive and ICT-enabled SMEs. The initiative also sought to foster the international dissemination of successful practices on business incubation, especially on the effective use of ICT and e-services, to achieve greater efficiencies, higher productivity and better performance in these programs and their client small-sized enterprises (Hernandez, 2003).

E-Readiness of SMEs in the Caribbean

In the Caribbean, several SMEs have grown and expanded into international markets through the use of ICTs by raising the quality of their products, and improving on customer services, information access and flow. In Jamaica, for example, there is a growth in the use of ICT in the SME sector, which has been attributed to increased awareness and the dramatic drop in the global costs of hardware and software. The Jamaican experience emphasizes the importance of differentiation and access to finance as factors that could encourage SMEs in competition and expansion into international markets (Southwood, 2004). Perhaps recognizing this, the government of Trinidad and Tobago released a national ICT plan in 2004 that sought to encourage SMEs to transform into e-businesses in order to compete in arenas that were traditionally only open to larger players (Houng, 2004).

E-Readiness of SMEs in Africa

The e-readiness status of SMEs in Africa is still largely un-researched. However, it would seem as though some strides are being made in a few countries. Within the SADC states, for example, there is general acknowledgement that e-commerce presents a major opportunity for SMEs to compensate for the countries' traditional lack of access to national and international markets. Although small-size entrepreneurship is widespread throughout the SADC region and its potential for improving local economies is well recognised, the SMEs are less able to manage and sustain the costs involved in adopting and investing in new technologies. There is therefore a strong rationale for getting governments in this region to address the problems that impede SMEs from adopting and using ICT and electronic commerce (SADC, 1999:35).

Within certain individual countries, such as South Africa, annual reviews of the e-readiness of SMEs have regularly been commissioned. The Sunday Times' Business Times supplement - a reputable weekly newspaper – has, since 2003, commissioned e-readiness assessments of SMEs in the country. In one such study, Gordon (2003) found that a number of companies in the SME sector were fairly well plugged into technology. During 2003, e-readiness assessments of SMEs in South Africa showed that:

- 69% of SMEs had three or more PCs
- More than 11% had none
- 87% of those companies with PCs did not use the Internet to buy products
- 11% used the Internet intermittently.

The study largely concluded that there was huge growth potential for the Internet and networking technologies within the SME sector in South Africa. The study revealed that in the micro-enterprise sector, for example:

- 84% of employees had Internet access
- 84% had access in the very small category
- 89% had access in the small category
- 93% had access in the medium category
- 30% of the micro-organisations had websites
- 5% of very small companies had web presence
- 60% in the medium category had web presence.

Gordon (2003) noted that 74% of the companies believed that the Internet held future possibilities that they were yet to exploit. And on the whole, South Africa had more SMEs during 2003 reported to be using ICTs compared to the number of businesses in Mauritius that were connected to the Internet (Aubeelack, 2004:16-17). South Africa has a strategy known as Godisa aimed at nurturing innovation and incubating successful technology-based enterprises. It aims to stimulate economic growth and development as well as enable long term employment creation through technological innovation and the enhancement of productivity, sustainability and international competitiveness (Daniels, 2004:26). With the growing use of ICTs amongst SMEs in South Africa, companies offering new services have emerged. For example, Netsurit is an IT consultancy that specialises in the maintenance and service of SME networks (Netsurit, 2003). Netsurit provides SMEs with services such as Internet technology, e-commerce, e-privacy, e-services, portals, e-business and security, e-marketing, advertising, and software.

The Bodibeng technology incubator of South Africa focuses on ICTs and electronic technology, and supports hi-tech entrepreneurs in starting and growing their businesses by offering networking, mentoring and coaching. South Africa's other ICT initiatives developed to support SMEs include Softstart and the KZN (KwaZulu Natal) Innovation Support Centre in Durban, respectively. Softstart aims to assist innovative, early stage software entrepreneurs by providing mentoring and support for the growth of sustainable business. The KZN Innovation Support Centre provides technical and business support that enables innovators in the electronic, information and communication technology sectors to sell their ideas on the market. It provides business, research, technical, management and marketing skills to support innovation (Daniels, 2004:29).

Further north of Africa, the Egyptian private sector and its government have played a key role in investing in ICTs in the SME sub-sector in order to ensure their sustainability and improve their profitability (Maksoud and Youseff, 2003).

PROPELLING SMEs TO AN E-READINESS STATUS

There are several drivers that have the potential to propel a firm, community or nation to a high e-readiness status. In order for e-readiness to be achieved, a country needs robust infrastructure that consists of, among other components: telecommunications services, IT hardware and software, an adequate power supply, high teledensity, Internet access, high bandwidth, and good quality networks. A strong infrastructure in terms of telephone penetration and bandwidth would define the access points for services offered by governments and businesses. E-readiness also has legal oriented issues, such as national IT policies, cyber laws, consumer protection, security of transactions, and policies on the use of IT, that should be addressed (Department of IT eTechnology Group -India, 2003).

Consulting and Audit Canada (2004:34) indicated that the evolution of e-readiness is the spur behind a knowledge-based society, wherein ICT plays an important role in accelerating the creation and sharing of knowledge. To share information, four components are important, namely ICT infrastructure, language tools, structure, and content. ICT infrastructure refers to the physical components necessary to access the Internet, such as computing devices or the hardware required to store bits of data. ICT language tools and services are the means through which ideas are transformed into digital content. They make it possible to codify information, be it text, video or graphics, so that it may be transmitted along electronic networks. Structure generally refers to the conventions that have been adopted for interpreting codified language. It enables data to be organized and arranged in a way that is comprehensible to the users of ICT. The content is also important, because it is critical to the use of the infrastructure. It is therefore important for e-readiness assessments to take cognisance of these components because they are critical in the creation and sharing of information.

Investment in establishing high e-readiness capability and the ongoing maintenance costs can be justified if they are more than offset by the benefits. Such benefits would include (Agriculture and Food Canada, 2001):

- Avoidance of market erosion
- Market expansion
- Decreased inventories
- Decreased transaction costs.

The costs in general refer to both the cost of introducing e-readiness investment and the maintenance costs. The costs of introducing e-readiness would vary according to the scale of economies as determined by the number of transactions and the size of the supplier/customer network. The unit costs would usually vary inversely to the activity of transactions and the size of the network. It is necessary to assess the e-readiness of the firm's own supplier-customer network before a decision to adopt is made, because it is possible that although the firm may have achieved a certain level of e-readiness to transact business over the Internet, its suppliers and customers might not be ready to do so. Furthermore, the size of the firm and the characteristics of its management influence the firm's degree of e-readiness. It is also possible that some SMEs might not be logical candidates for e-commerce (e.g. levels 6 and 7 based on Agriculture and Food Canada's model of e-readiness) at a particular point in time, whereas other larger firms might find it critical to do so.

E-readiness adoption decisions are like any other capital expenditure decision, and each firm decides the level at which the benefits outweigh the costs before moving forward to ensure that it reaches and maintains or surpasses that level. As already pointed out, a firm's specific variables, such as the management's propensity to change and the education level of management (especially the chief executive), are important in determining the degree to which an individual firm adopts the Internet as its principal means of doing business. The priorities of different firms will inevitably differ as far as e-readiness adoption decisions are concerned. For example, the adoption of e-commerce might not necessarily be appropriate for all firms in an industry, even if on average there are benefits to be gained from it. Take for example, an SME that is owned by one person and is local; buys a single homogenous product from one long-term supplier with no expected changes to that purchase; and sells this exclusively to a large long-term buyer with standard monthly invoices and payments, with no major expected changes in this relationship. Such a firm would have no immediate requirement for an expensive e-commerce platform or even a website (Agriculture and Food Canada, 2001).

The World Summit on Information Society, through its Declaration of Principles, implicitly outlined what governments, civil society, NGOs, the Media and the private sector need to do in order to participate in the digital economy. In particular, the Declaration of Principles (WSIS, 2003) called on governments to: build institutional capacities to collect, organize, store, and share information and knowledge; stimulate the creation, processing, and dissemination of information to all people; conserve local content; enable access for all people to information through the use of ICTs; facilitate speed and ubiquity of information flows through ICTs; enhance information sharing within countries and across the globe; use ICTs to improve productivity and the quality of life; create, receive, share and utilize information in any media, regardless of frontiers; develop high quality ICT networks; enhance effective legal and regulatory frameworks for unhindered access by individuals to communication media and information sources; build secure networks to enhance information and establish trust in the use of ICTs; protect privacy through confidentiality; provide human resource development strategies; develop human capacity to exploit the benefits of ICTs; build public awareness on the capabilities of ICTs; put in place education and training programmes on ICTs; put in place continuous training programmes for the benefit of all; facilitate the free flow of information and ideas from a diversity of sources; and enhance linguistic diversity and multilingualism.

The Declaration of Principles also called for creating conducive conditions for the production, processing, and dissemination of local content; eliminating illiteracy and enhancing ICT literacy; bridging the digital divide; enhancing universal access through the deployment of affordable ICTs, including radio and television; improving connectivity; using open standards to enhance affordable access to ICTs and interoperability; providing a supportive and predictable policy, legal and regulatory framework; and enhancing innovation and competition. Lastly, the Declaration of Principles proposed:

- Enhancing the use of ICTs in people's everyday lives
- Providing technical assistance and support with regard to ICTs
- Recognizing the benefits of ICT applications
- Making available appropriate electrical power sources.

SMEs that have a high prevalence of best practices in the use of ICT are more likely to be productive than those that do not. Such enterprises recognise the need to manage information through formal strategies and structures. Businesses that are able to effectively employ important information and communication technologies also find more sophisticated and efficient ways to manage their external relationships and communications. Pervasive ICT usage in organizations helps them formulate a critical mass of electronic transactions that support a networked economy, both in terms of the network size and the demand for associated goods, services, and labour policy reforms. Businesses that invest more heavily in computer technology are more likely to be staffed with people that are trained to take advantage of it. When skilled people have access to technology, their productivity is significantly enhanced, and when this is the case across the board, the economy grows by leaps and bounds.

Small-sized enterprises in developing countries can enhance their competitiveness by using ICTs to participate in new export sectors, streamline business activities, and link more effectively with trade development partners. Competitive advantages can also be achieved if SMEs go beyond mere Internet access to developing e-business capacity and investing in strategic alliances. Moreover, SMEs stand to compete well in the international market if they have the human resource capacity (technical and managerial) to harness ICTs effectively for daily use. They also need to have access to affordable ICTs such as personal computers, Internet applications and related software.

Rao (2003) proposes five recommendations that can help SMEs join the Internet and wireless world, namely:

- Focusing on in-house applications, ranging from anti-virus protection and website design to knowledge management and business intelligence
- Developing in-house competency to harness tools for e-business
- Combining physical and electronic channels
- Developing alliance strategies
- Improving their web presence.

Public policy and legislation can help or hinder the development of a mature networked economy. A favourable climate can be created by an appropriate legislative and regulatory regime that encourages communities, organisations and individuals to invest in and use ICTs. Important areas, such as Internet access and availability, the use of ICT in schools, and the growth of e-commerce; are all influenced by public policy and legal frameworks. Furthermore, any country wishing to participate actively in the modern connected world should have in place policies and legislations that ensure that electronic commercial transactions can be enforced and electronic documents and signatures authenticated. Citizens must be able to trust the electronic environment because issues of transactional security, privacy, and data integrity have been addressed. The potential for fraud and the cross-border transmission of objectionable content as well as new criminal behaviour must also be addressed and enforcement arrangements agreed upon.

The legislative and policy regimes surrounding the provision of telecommunications services are a key element in determining whether a society is in a strong position to take advantage of e-communication possibilities. This element of e-readiness can influence the strength, capacity and flexibility of the infrastructure; the vibrancy of the competitive environment; the choices available to consumers; the likelihood of investment in e-commerce; and the stability of a legal structure that promotes innovation and attracts foreign investment. A high degree of e-readiness in this area is showcased by modern, flexible legislation and a competent credible regulator armed with a strong mandate and a range of powers that foster and enforce competition and a fair regulatory playing field. The favourable climate that legislation can create for Internet use and e-commerce encourages communities, organizations and individuals to invest and use information and communication technologies.

In order for SMEs to effectively compete in an international market, it is important for governments to facilitate access to credit and finance. Governments should identify various business models that SMEs can afford, manage and maintain to participate in international markets the way businesses in the US, Britain and other developed countries do. Governments should also provide a one-stop portal that supplies updated information on business and also provides a link to international marketing initiatives. Affordable information and communication technology and services are also part of the necessary preconditions for participation in the networked world. Due to the high costs of Internet connectivity, a funding mechanism, perhaps in the form of a SMEs access fund, is necessary to enhance connectivity. Sachs (2003) notes that in communities where the sum of ISP and telephony fees is prohibitively high, a disincentive to use the network exists and access is curtailed. The provision of tiered pricing packages could improve affordability for many subscribers by allowing them to purchase only what they need. Governments need to strive to ensure that PC technology is accessible to those who want it by providing grants, subsidies and corporate donations, especially seeing as this has proved effective in countries such as Estonia, Malaysia and Mauritius (World Economic Forum, 2003). Governments can also provide incentives in the form of effective business support services to facilitate the development of appropriate partnership arrangements; avail long term finance; and create an appropriate and effective legal and regulatory framework that safeguards intellectual property rights, brand protection and local contract enforcement.

CONCLUSION

Internationally, SMEs in developed economies, such as those found in North America, Europe and parts of Asia, have attained high levels of e-readiness compared to those in developing countries. SMEs in developing economies face myriad problems, including poor awareness of the benefits of ICTs; lack of access to credit; inadequate information to make informed decisions; lack of expertise; high cost of access to ICT infrastructure; high taxation, etc. Propelling SMEs to high e-readiness statuses requires, among other actions, the provision of robust infrastructure (e.g. Internet access, quality networks, enabling policies, affordable hardware, etc).

Despite the challenges facing SMEs, they are increasingly integrating ICTs into their business transactions, especially in terms of the use of the Internet and the World Wide Web. There is a distinctive pattern of progression in the adoption of the Internet among SMEs, from first using simple static websites, to the interactive stage, where ordering and payment can be made online. Publicly owned SMEs are more likely to adopt ICTs, followed by private, non-family owned enterprises, and lastly family owned enterprises. A number of factors motivate SMEs to adopt ICTs in their business transactions, key among which are the perceived benefits (e.g. ease of communication, advertising, creating portals, online banking, e-procurement and e-payments). Other factors that influence the adoption of ICTs by SMEs include the educational level of the CEO, sophistication of the intended online transactions, and the costs of technology.

Governments and the managerial authorities in SMEs must provide leadership by enhancing national policy frameworks (the role of government) and promoting organisational readiness (the role of SME managers) to create an enabling electronic business environment where enterprises can grow and partake in the global digital economy. Governments must also ensure affordable access to infrastructure (e.g. telecommunications, computers, etc) and facilitate an enabling legal environment that supports electronic necessities like digital signatures and encryption. At organisational level, management authorities must ensure that requisite infrastructure and adequate skills are available to enable their enterprises to transact business electronically.

REFERENCES

Agriculture and Food Canada. (2001). *The e-readiness of Canadian Agriculture and Food Canada SMEs*. Retrieved on March 21, 2004, from http://www.agr.gc.ca/misb/fb/food/ecommerce/eready/introduction_e.html

Aubeelack, P. (2004, April 14-16). *E-government in Mauritius*. Paper presented at the SADC E-Government Workshop, Gaborone, Botswana (pp. 1-37).

Borg, A. (2004, April 14-16). *Organisational and cultural transformation: The Malta experience*. Paper presented at the SADC E-Government Workshop, Gaborone, Botswana (pp. 4-37).

Commission of the European Communities. (2002, June 21-22). *E-Europe 2005: An information society for all*. An action plan presented at the Seville European Council, Lisbon (pp. 1-23).

Consulting and Audit Canada. (2004). Transformation for the digital age: The development of Botswana's national ICT policy (pp. 1-81). *Preliminary Report for Consultancy for the Formation of a National Information and Communication Technology Policy Technical Bid.* Toronto: Government of Canada.

Daniels, G. (2004). All systems go for SME growth. *Enterprise*, 23-28.

Dejonckheere, J., Ramioul, M., & Van Hootegem, G. (2003). *Is small finally becoming beautiful? Small and medium-sized enterprises in the new economy.* Brighton: Institute for Employment Studies.

Department of IT eTechnology Group (India). (2003). *Assessment of central ministries and departments. E-governance readiness assessment 2003, draft report* (p. 48).

Gordon, G. (2003, February 16). SME survey. *Sunday Times Business Times, 15.*

Gungen, T. (2003). *E-Finance for SMEs in Turkey: Yapi Kredi Bank.* Retrieved on November 27, 2003, from http://r0.unctad.org/ecommerce/event_docs/xmefinoct2001/gungen.pdf

Hernandez, R. (2003). *RFP: InfoDev incubator initiative.* Retrieved on December 5, 2003, from http://www.infoDev.org

Houng, M. H. (2004). Trinidad and Tobago national information and communication technology plan. *INASP Newsletter.* Retrieved on April 22, 2004, from http://www.nict.gov.tt

Maksoud, S. S., & Youseff, M. A. A. (2003). *Information and communication technology for small and medium enterprises in Egypt* (pp. 1-15). Cairo: SME Development Unit-Ministry of Foreign Trade.

Minton, S. (2003). Nordic nations still top information index. *The World Paper.* Retrieved on August 13, 2004, from http://www.worldpaper.com/2003/oct02/isi1.html

Netsurit. (2003). *Outsourced IT management.* Retrieved on November 28, 2003, from http://www.netsurit.com/content/AboutUs.asp

Press Release. (2002). *E-Readiness assessment sees RP potential for technological growth.* Retrieved on November 27, 2003, from http://www.ifodev.org/news/announce/020604.htm

PRWeb. (2003). *78% of Jordanian SMEs utilize computers.* Retrieved November on 21, 2003, from http://www.prweb.com/releases/2003/3prweb58999.htm

Ramayah, T., M. Jantan, R. Mohd Roslin, and R. Siron (2003). Technology readiness of owners/managers of SME's. International Journal of Knowledge Culture Change Manage, 3, 475–486.

Rao, M. (2003). Checklist for national e-readiness. International Trade Forum, 3, 10.

Sachs, J. D. (2003). *Readiness for networked world.* Retrieved on November 28, 2003, from http://cyber.law.harvard.edu/readinessguide/intro.html

SADC. (1999). *SADC in the next millennium: The opportunities and challenges of information technology* (pp. 1-64). Gaborone: SADC Secretariat.

Selhofer, H. (2003, June 3). Indicators for the new e-business environment: Challenges for statistics and research. *A Presentation at the DEEDs 5th Policy Group Meeting,* Brussels.

Southwood, R. (2004a). The impact of ICT on SMEs-a motor for future economic growth in hard-pressed times. *Balancing Act News Update, 192.* Retrieved on January 26, 2004, from http://www.balancingact-africa.com/news/current1.html

Southwood, R. (2004b). Kenya airways e-commerce site shows it can be done but Africa's banks keep up? *Balance Act News Update, 204.* Retrieved on April 19, 2004, from http://www.balancingact-africa.com

Trend Chart on Innovation. (2000*). Innovation and SME programme. European commission (directorate general enterprise)*. Retrieved on April 17, 2004, from http://www.cordis.lu/trendchart

World Economic Forum. (2002). *Southern African development community-world economic forum consultation report on e-readiness* (pp. 1-43). Geneva: WEF.

World Economic Forum. (2003). *Global information technology report.* Retrieved on August 24, 2004, from http://www.weforum.org/site/homepublic.nsf/Content/Global+Competitiveness+Programme%5CGlobal+Information+Technology+Report

WSIS. (2003). *Draft declaration of principles.* Geneva: WSIS Civil Society Working Group. Retrieved on November 7, 2004, from http://www.wsis-pct-org/prepcom2-declaration.htmlf

Section 4
Digital Economy Applications

This section consists of Chapters 9 and 10 (E-Government Readiness and E-Business and E-Commerce, respectively). Section 4 expands on the previous section by examining how e-readiness is critical in the implementation of e-applications such as e-government and e-business/commerce. Governments have a major role to play in providing an environment that is conducive for electronic business transactions by laying requisite infrastructure and enhancing a legal and policy-driven framework. Chapter 9 therefore covers the nexus between e-government and the digital economy, applications of e-government in society, global e-government trends, global e-government rankings, e-government survey methodologies, benefits of e-government systems, challenges of implementing e-government systems, and the policy and managerial implications of implementing e-government systems. Chapter 10 covers e-commerce and e-business as major applications of the digital economy, the development of e-commerce, types of e-commerce, determinants of e-commerce adoption by SMEs, e-government/e-commerce nexus, e-commerce implementation in e-government settings, e-commerce readiness (connectivity, business environment, legal and regulatory environment, social/cultural infrastructure, etc.), methodologies for assessing e-commerce readiness, and e-commerce challenges. The chapter aims to provide information that can prepare SMEs for an e-commerce environment. It also aims to show how e-commerce may offer SMEs ways of finding new customers and suppliers, especially in international and regional markets which they have not easily been able to reach before.

Chapter 9
E-Government Readiness

INTRODUCTION

The United Nations (2008) believes that in a world characterised by rapid changes driven by globalization and the emerging knowledge-based economy, e-government provides the means to achieve maximum cost savings and improved service delivery to both citizens and businesses. In short, the digital economy and e-government can be said to be inextricably intertwined. Global e-commerce growth and expansion in the private sector are linked to an online population that is projected to reach some 1.8 billion by 2010 (United Nations, 2008). Most of this population will be using e-government infrastructure to go online. Moreover, the widening scope of digital technologies through e-government means that few (if any) industries are exempt from some degree of transformation and electronic commerce. The UN 2008 E-government Readiness Report showed that in the United States alone, online retail sales were expected to reach nearly \$120 billion by 2008 (United Nations,

DOI: 10.4018/978-1-60566-420-0.ch009

2008). In less developed countries, similar trends are apparent, with Argentina, for example, having recorded a more than 100% increase in e-commerce levels from 2005 to 2006 - more than $3.3 billion online transactions were recorded in the latter year. Much of e-government reflects private sector activity, which has both encouraged and pressured public sector organizations to act in a similar manner. Fiscal constraints imposed by a quasi-competitive system of global investors, as well as a strategic desire to generate cost savings and reallocate spending to new priorities, make the nexus between technology management and efficiency a central concern in government (United Nations, 2008).

E-government is being applied in various ways in society. Some of the applications include e-administration, e-business, e-services, and e-society. These applications define the level of interactions that define relationships between the government and their respective stakeholders. E-administration applications arise from interactions within and between government agencies, as well as between different spheres of government at local and provincial levels. These interactions are sometimes referred to as government-to-government transactions, or simply G2G. Such applications seek to improve the operational efficiency and effectiveness of government. E-administration applications provide facilities that enable electronic communication and the sharing of information and knowledge. They also permit simultaneous access to information, thus shortening the bureaucratic red tape associated with the transmission of information.

E-business applications aim to improve interactions and relationships between the government and its customers (e.g. suppliers) who are usually members of the business community. This interaction is generally referred to as government to business or simply G2B. However, the interactions and relationships could be with business entities, citizens or any other legal entity in which the government has a business interest. E-business applications digitize most government procurement and disposal of assets, thus cutting the red tape, middlemen and time required to process applications or obtain information, and hence reducing operational costs, unnecessary delays, paperwork and redundant data capture. Consequently, these applications can improve the business and economic competitiveness of a country. By permitting direct interaction between the respective parties, e-business cuts out the middleman and transfers the resulting cost benefits to the buyers. This leads to the increased affordability of goods and services. In addition to speeding up processing time, e-business creates a more transparent business environment, thus leading to better and timelier decisions and actions. It also leverages new technologies for better communication between the respective parties (Oyomno, 2003).

Electronic services (e-services) enable interactions between the government and citizens, through which the latter gain access to a wide range of public services.

These applications are referred to as government-to-citizen, or G2C. However, these interactions and relationships may not be limited to citizens only. They may also include non-citizens and other legal entities that the government interacts with in order to deliver public services. E-services enable all branches and levels of government to function as a single coordinated entity, thus expanding a government's availability and accessibility, and moving government in the direction of 'anytime, anywhere and by any means'. The goal of 'anytime' availability is basically the procurement of around-the-clock service. E-services provide further choices by allowing consumers to use a variety of technologies to gain access, such as the telephone, fax, e-mail, kiosks, face-to-face interaction, and more. Typical public services that can be delivered electronically include permits and registration, directory services, public information and records, social grants, school and library information, and revenue collection.

E-society applications, on the other hand, enable government to engage with the collective membership of communities and societies that comprise the nation. These applications create convenient channels for consultations and dialogues between the government and its stakeholders, thus facilitating interactions and relationships. These applications are generally referred to as government-to-society, or simply as G2S. E-society applications give communities a collective voice in their dealings with the government in order to enhance the quality and effectiveness of communication, and encourage active and productive participation in the governance process. E-society applications' capacity to enable interactive participation and multiple consultations between the government and all its stakeholders is sometimes referred to as 'e-democracy' (Oyomno, 2003).

GLOBAL E-GOVERNMENT TRENDS

E-government refers to the application of information and communication technology within public administration to optimise its internal and external functions, [thereby providing] government, the citizens and businesses with a set of tools that can potentially transform the way in which interactions take place, services are delivered, knowledge is used, policies are developed and implemented, citizens participate in governance, public administration is reformed, and good governance goals are met (UNDESA, 2006). Several concepts revolve around e-government. For instance, the concepts of e-government, e-governance and connected governance are sometimes used interchangeably. However, *e-government* connotes the application of ICTs to facilitate social governance processes or objectives, such as information for political participation, consultation and consensus-seeking among governments, public servants, politicians and citizens (Sheridan and Riley, 2006). *E-governance* is an advanced level of e-government, where citizen-government engagement takes place electronically.

The term *connected or networked governance* is a deepened form of e-governance and refers to governmental collective action to advance the public good by engaging with the creative efforts of all segments of society. It is about influencing the strategic actions of other stakeholders (United Nations, 2008). Connected governance efforts are also aimed at improving cooperation between government agencies; allowing for enhanced, active and effective consultation and engagement with citizens; and facilitating the greater involvement of various stakeholders, both regionally and internationally. Connected governance provides better organized, aligned and often integrated information flows, new transactional capacities, new mechanisms for feed-back and consultation, and more participative forms of democracy. For those engaged in the management and delivery of public administration, connected governance is about driving down costs and improving the effectiveness and efficiency of back office functions. For those working at trans-national level, connected governance is about removing barriers to international cooperation and development and creating a global agenda of connected governance (United Nations, 2008).

The other evolving concept around e-government is what has come to be known as *e-government-as-a-whole,* which focuses on the provision of services at the front-end, supported by integration, consolidation and innovation in back-end processes and systems to achieve maximum cost savings and improved service delivery. A related but also new concept, the *whole-of-government,* refers to "public service agencies working across portfolio boundaries to achieve a shared goal and an integrated government response to particular issues" (United Nations, 2008). The distinguishing characteristic of the 'whole-of-government' approach is that government agencies and organizations share objectives across organizational boundaries, as opposed to working solely within an organization. It encompasses the design and delivery of a wide variety of policies, programmes and services that cut across organizational boundaries. The 'whole-of-government' concept is a holistic approach to ICT-enabled public sector governance (United Nations, 2008).

All emerging forms of e-government aim to enhance service delivery by simplifying bureaucratic procedures; facilitating efficiency and transparency; improving information sharing and the innovation of service; and increasing the level of citizen empowerment. Because of the benefits that can be accrued from the implementation of e-government, the UN and the World Bank have adopted e-government as a developmental instrument (Cloate, 2007). E-government is a trend emerging around the world aimed at reforming the public sector through the involvement of citizens, the private sector, and non governmental organisations that both want and need governments to provide quality services. These constituents are demanding from government top performance, efficiency, proper accountability, public trust, and enhanced service delivery. The success of government leaders is increasingly being measured according to the benefits they are creating for their constituents. In response, governments are introducing innovations in their organizational structures, practices and capacities; in the ways they mobilize, deploy and utilize human capital and information; and in the technological and financial resources they use to improve service delivery to their citizens.

E-government is perceived as a panacea to the deficiencies of traditional forms of government, wherein citizens physically go to government offices to apply for passports, birth certificates or death certificates; file tax returns, etc.; with the consequent delays that arise out of long queues, lost files or the absence of relevant officials. E-government enhances the overall service/product quality when compared to traditional forms of government where there is an unnecessary level of red-tape, bureaucracy and restrictive procedures and processes. E-government ensures that government processes and services observe the law and maintain their integrity in satisfying citizen needs through the delivery of relevant, value-added and high-quality services. These services are expected to enhance accountability and integrity in government; offer the prospect of cheaper and more effective management and processing of information; facilitate the free flow of information between departments; enhance transparency, especially with regard to the procurement of services; provide opportunities to work in partnership with the private sector; and enable citizens to participate directly in governance, especially in influencing policy decisions. E-government also enables governments to offer services and information through new media like the Internet; improves communication between different parts/levels of government so that people do not have to be asked repeatedly for the same information by different service providers; provides staff in offices with better access to information so that they can deal with members of the public more efficiently and more helpfully; makes it easier for different parts of government to work in partnership with central government, local authorities, the post office, private sector companies, etc; and transforms government into a learning organization by improving access to organizational information (Chadwick and May, 2003).

The impetus for thinking about more online dimensions to public sector operations came into being in the 90s, when the mainstream advent of the Internet began to translate into dramatic declines in the cost of both communicating and processing information. Networking and the more transformational potential of the Internet also promised more in terms of rethinking how and why governments function. The initial face of e-government then was the website, initially static in form but soon enriched and transformed into a 'portal' with online functionality and multiple applications. During the 90s, as countries and other jurisdictions began to develop their web presence, it became apparent that the mere online listing of government departments and agencies would not be the most effective way to develop more transactional and interactive capacities in an efficient and effective manner. According to Remmen (2004), three different notions of e-government emerged during this period, namely efficiency (i.e. cost reductions); public service (better quality; easier, 24/7 access; and new services); and democracy (participation and interactive dialogue).

E-government is emerging as a powerful tool for the support of business enterprises, especially SMEs (notwithstanding the challenges they have to overcome), which have traditionally been marginalized from mainstream business by large businesses. Adeshara et al. (2004) studied SMEs' acceptance of e-government services in the UK, and found that most SMEs used e-government services for their business needs. Moreover, most SMEs appreciated the value of e-government in enhancing their business competitiveness locally and internationally. As far as the dependency on e-government services is concerned, 9% of the companies were found to be highly dependent on e-government services. In terms of awareness, 38% of the companies became aware of the government websites and e-government services through a search engine, and 24% were linked to e-government services through other websites. 14% were informed about e-government services by word of mouth, 6% through post, and 6% through television.

With regard to the problems that affect the propensity of SMEs' use of e-government services, 28% of the companies expressed concern about the security of transactions, 21% were wary of privacy violations, 21% were concerned about how quickly and easily they could use the services, and 20% were concerned about the cost of services. Overall, the study concluded that there is a demand for e-government services amongst the UK's SMEs, but the percentage of companies using them is still moderate. The most frequently used e-government services consisted of online legal advice, self assessment tax returns, training, information on how to write a business plan, and grants and support schemes. Policy and managerial decisions should therefore aim to create awareness; enhance education and training; strengthen the legal and regulatory framework of e-transactions; and enable affordable access to requisite infrastructure.

GLOBAL E-GOVERNMENT READINESS RANKINGS

The past five years have witnessed global efforts to assess governments' preparedness to participate in the networked world. Several surveys on e-readiness in general and e-government readiness in particular have placed most countries in the developing world, especially those in Africa, in the bottom tier of the world's League of Nations. The 2008 e-government readiness survey by the United Nations (2008) shows that there are large differences between the five regions of the world in terms of e-government readiness indices, with Europe (0.6490) having a clear advantage over the other regions, followed by the Americas (0.4936), Asia (0.4470), Oceania (0.4338) and Africa (0.2739). The poor performance of developing countries is apparent. During the 2008 survey, European countries made up 70% of the top 35 countries, while Asian countries made up 20% of the same. The report notes that a large part of the success of European countries can be attributed to their investment in infrastructure and connectivity, most notably in broadband infrastructure. During the 2008 survey, there were no countries in the top 35 from the African, Caribbean, Central American, Central Asian, South American and Southern Asian regions. From Africa, for example, South Africa took 61st position, followed by Mauritius (63rd), the Seychelles (69th), and Nigeria (79th).

A similar survey of 191 UN member states for the E-government Readiness Index of 2005 showed that while developed and transitional economies had made remarkable strides in electronic government, African states' performance remained poor. The 2005 UN Report (Holzer and Kim, 2005) noted that despite some progress with regard to e-government implementation, a serious access divide existed across the world between developed and developing countries. Of particular concern were countries belonging to the regions of Africa, South Asia and Central Asia. These countries showed little relative progress in 2005 with respect to outreach and access to citizens. A study conducted by the United Nations and the American Society for Public Administration in 2005 on the global digital government of 98 municipalities, found that only the city of Cape Town (South Africa) from Africa made it to 31st place (Holzer and Kim, 2005). The evaluation focused on, among other variables, the delivery of public services; digital democracy (citizen participation in government); and the security, usability, and content of websites.

E-GOVERNMENT READINESS SURVEY METHODOLOGIES

The E-government Readiness surveys are generally based on a holistic view of development that incorporates human capacity, infrastructure development and access to information and knowledge; 'government to citizen' (G2C) and 'government to business' (G2B) transactions; and e-government leadership. Other considerations include methods of delivery (such as the Internet and cellular phones, as well as access to PCs), and the capacity of a country to absorb content and services. However, the surveys' focus may vary from year to year depending on various critical issues of the time. Measures and indicators of the quality of web content include interactivity, dynamic versus static sites, community-centric or citizen-centric sites, regularly updated content, and downloadable content, to name a few (United Nations Division for Public Economics and Public Administration, 2002).

The E-government Readiness Index is a composite index that pulls together the web measure index, the telecommunications infrastructure index, the human capital index and the e-participation index. The web measure index of 2008, for example, is based on a five-stage model that builds on the previous levels of sophistication of a member state's online presence. As a country migrates upwards through the various stages, it is ranked higher according to the web measure index. The web measure survey assessments of 2008 were based on a questionnaire, which allocated a binary value to the indicator based on the presence or absence of specific electronic facilities or services. The primary site was the national portal or the official government home page of the member states. Where no official portals were available, other governmental sites were assessed. The phases of the web measure index are those that have been discussed already, namely emerging, enhanced, interactive, transactional and connected (United Nations, 2008). As countries move upwards in their attempts to reach the stage of a connected government, they pass through many thresholds in terms of infrastructure development, content delivery, business re-engineering, data management, security, and customer management. The web measure index provides member states with a comparative ranking on their ability to deliver online services to their citizens (United Nations, 2008).

The telecommunications infrastructure index, on the other hand, is a composite index of five primary indices that relate to a country's infrastructure capacity based on the delivery of e-government services. These are (United Nations, 2008):

- Internet Users /100 persons
- PCs /100 persons
- Main Telephones Lines /100 persons
- Cellular telephones /100 persons
- Broadband /100 persons.

Each index represents 20% of the overall telecommunications infrastructure index. The source of the telecommunications infrastructure data obtained for each member state is the International Telecommunication Union (ITU). Constructing five separate indices for the indicators standardizes the data across countries (United Nations, 2008). The human capital index is a composite of the adult literacy rate and the combined primary, secondary and tertiary gross enrolment ratio, with two thirds weight given to the adult literacy rate and one third to the gross enrolment ratio. The data for the adult literacy rate and the gross enrolment ratio is drawn primarily from the United Nations Educational Scientific and Cultural Organization (UNESCO).

The e-participation index measures the extent to which governments are able to establish more transparency by allowing citizens to use new channels of influence that reduce barriers to public participation in policymaking. Moreover, e-participation looks at how governments are able to create an environment that allows citizens to voice their views online and more importantly, to create a feedback mechanism showing citizens that their views are taken seriously. The e-participation index assesses the quality and usefulness of the information and services provided by a country in order to engage its citizens in public policy through information and communication technologies. Member states' levels of e-participation are assessed according to their institutional capacity, leadership role, and willingness to engage with their citizens by supporting and marketing participatory decision-making when developing public policies; and the structures put in place to facilitate citizens' access to public policy dialogue (United Nations, 2008).

In calculating the e-participation index, a total of 21 citizens' informative and participatory services and facilities were assessed across 189 countries in the 2008 survey. This was only in instances where these services and facilities were online and where data was available. Questions were grouped under three categories, namely e-information, e-consultation and e-decision making. Each country was assessed on a scale of 0-4. The index was constructed by standardizing the scores.

E-information refers to a government website that offers information on the list of elected officials; government structures, policies and programmes; points of contact; budget; laws and regulations; and other information of public interest. Information is disseminated using a number of online tools such as community networks, blogs, web forums, text messages (micro democracy), newsgroups, and e-mail lists. E-consultation, on the other hand, consists of a government website that provides the tools necessary for e-consultation and allows citizens to set the agenda for debates through e-petitioning. The government ensures that its elected officials have a website to communicate directly with their constituents. It maintains an archive of their discussions and provides feedback to citizens. Lastly, e-decision refers to the government's willingness to take into account the e-inputs of citizens in their decision-making processes. The government informs its citizens on what decisions have been taken based on the consultation process (United Nations, 2008).

BENEFITS OF IMPLEMENTING E-GOVERNMENT SYSTEMS

In general, most governments across the world are moving to implement public sector reforms, which in many ways include ushering in a period of e-governance for various reasons. For one, there is increasing access to the web at home, work, school, and in cyber cafes. Governments also recognize that counter transactions often consume more staff time and more paper supplies than electronic transactions. Furthermore, as the online population grows, customers expect governments to have an Internet presence and offer online transactions (The New York State Archives, 2005).

Through e-governance, citizens become better informed and more involved in the activities of government. Essentially, e-governance in many ways compels governments to be more accountable, as it integrates operations across jurisdictions while blurring departmental boundaries, resulting in increased productivity and efficiency (New York State Archives, 2005). Through e-government, it is possible to achieve greater convenience and revenue growth. E-governance also enhances democratic processes and strengthens support for public policies.

Foreign investors look at different locations and their resources, skills, technology and infrastructure before making investment decisions. They also look at how helpful and efficient the respective governments are, how easy it is to get things done (e.g. applying for trade licenses), and how efficient the legal systems are. If the government process is very slow and difficult, investors are pushed to reconsider their choice of location, as it may not be a good place to do business. E-governance can also help decrease the possibility of maladministration and corruption by taking the human factor out of many processes through automation, particularly in the fields of procurement and in revenue collecting activities such as taxation and customs duties.

CHALLENGES OF IMPLEMENTING E-GOVERNMENT SYSTEMS

The United Nations (2008) observes that in some instances, governments have spent vast amounts of money building online systems and products only to find that their citizens do not fully utilize them. A number of reasons could be behind this, such as the citizens' unwillingness or disinterest; misunderstanding the needs of the people;

inadequate infrastructure; inadequate delivery of services; poor access to useful and accurate content; language, social and cultural barriers; lack of trust; and/or lack of confidentiality. The UN report concludes that in order for e-government to be successful, people must be willing and have the trust and confidence to use the e-government services provided to them

Citizen-government engagement in e-government environments presupposes the ability of the citizens and their willingness to use technology. Yet in most countries, especially in the developing world, low overall literacy levels bar effective e-government implementation. This is further exacerbated by low Internet penetration in (most) developing countries. Africa, for example, has the lowest Internet penetration in the world at 3.6% (Onyancha, 2008 citing Internet World Statistics, 2007). E-government operations may also be undermined by citizens' lack of technological competence.

The interconnectedness of the e-government environment raises issues of trust, which must be guaranteed by government for citizens to accept and interact in an e-environment. Such issues include, for example, privacy, the security of network infrastructure, data protection, integrity of the content provided, continued availability of infrastructure, protection of citizens on issues of intellectual property (as government portals often link to other websites in businesses, academia, etc), consumer protection, availability of an enabling legal framework that enhances e-commerce and stems uncompetitive behaviour, freedom of access to information, and more ((Bridges.org, 2001).

Many countries fail to involve the public in electronically-enabled decision-making processes because politicians fear that e-democracy may result in a loss of power. A number of historical anecdotes litter the landscape of bad governance, thus giving credence to the doubts citizens may have about the sincerity of their governments to improve service delivery through e-environments. They include, for example, restrictions of access to government-held information; scuttling of the free press; human rights violations; lack of democracy; and endemic corruption in places such as Zimbabwe Kenya, Tanzania and Uganda (Githongo, 1998). In Egypt, for example, government officials beat Ahmed Maher Ibrahim, a 27-year-old civil engineer, for using Facebook to support calls for a general strike on May the 4th, 2008 - President Mubarak's 80th birthday. These kinds of actions do not inspire citizens to believe that governments can become any better in matters of transparency, accountability, freedom of access to information, etc; even with e-government implementation.

During the WSIS Summit in Tunis in 2005, civil society held demonstrations against the Tunisian and Moroccan governments, complaining about brutal state repression against free press and freedom of expression (World Information Society Report, 2006). In Kenya, the founding father of the nation, Mzee Jomo Kenyatta, was fond of stating: "Serikali ni Siri Kali" (translated literally from Swahili to mean 'government is top secret'). Governments that have traditionally restricted access to public information cannot be expected to embrace e-government whole-heartedly and avail information that would enable citizens to question their public policies, accountability and integrity. By and large, government systems in developing countries, especially in Africa, are characterised by bureaucracy and red tape, which in some countries are seen as a form of power and also a source of side income for corrupt civil servants. For instance, a person applying for a trading license may have to pay the person who dishes out the application forms, bribe the clerk to have their application stamped, and pay a string of other people just to get the application considered, let alone granted. E-governance breaks bureaucracy, which is not desirable to many [corrupt] government workers as this would directly affect their food-chain. Some governments fear the potential of the Internet and wish to control it and the companies that provide the services and products tied to it; and users fear the consequences of using it as a medium for openness and accountability (Ganesan, 2008).

Bagire (2005) observes that in many [developing countries] where websites exist, they contain little information of value to the public. In Africa for example, while information on "who is who" in the ministry, its organizational structure, and its mission will often be abundantly available, the average website will not have public service information, for instance on how to apply for a particular service, who the right officer/person to approach is, and where on the website to download and electronically submit application forms. The Ibrahim Index on African governance expressed concern about the difficulties that were faced in collecting secondary data from government websites, saying "not all African countries have websites and where they do, they may not post useful data" (Rotberg, 2007:20).

Furthermore, although most governments in the developing world are implementing e-government systems, little awareness is being created among the people. This is made worse by the lack of citizen involvement in the planning process. Xiong (2006), in the context of e-government in China, noted that if people have never heard of e-government, and have no interest, no equipment, and no knowledge or skills on how to access and use online government information and services, e-government is meaningless. Most developing countries also face problems related to the poor organization of knowledge in government portals, which deny citizens the benefits that could be gained from sound information and knowledge management practices. Identifying the right processes to capture, store and share knowledge is an essential aspect of knowledge management (United Nations, 2008).

In the developing world, where most people cannot effectively access the Internet because of factors already discussed, whatever e-government initiatives are made cannot go further if issues related to Internet access are not addressed. This situation is only made worse by power shortages. Cokayne (2008), citing the IMF in an article published in the Sunday Times (South Africa, May 19, 2008:18), warns that the power crisis in sub-Saharan Africa is a constraint to economic growth. The report indicates that the adverse effect of the power crisis affects productivity by 30 - 60%, ahead of red tape and corruption. It is estimated that more than 48 countries in Africa have suffered acute energy crises in recent years, and the overall generating capacity of these countries, at 63 gigawatts, is comparable to that of Spain (Cokayne, 2008).

WAY FORWARD IN IMPLEMENTING E-GOVERNMENT SYSTEMS

The phases of the e-government evolution act as both a benchmark and a framework for governments to enhance their e-government statuses and provide an enabling environment for the growth of business transactions. The first phase in the evolution is the *Infrastructure phase,* which requires information infrastructure - both within the public sector and across society at large – that can facilitate reliable and affordable Internet connectivity for all citizens, businesses and stakeholders in a given jurisdiction. This is followed by the *Integration phase,* which is basically concerned with leveraging new infrastructure within the public sector in order to better share information (both internally and externally) and bundle, integrate, and deliver services through more efficient and citizen-centric governance models using multiple delivery channels. The third and final phase is the *Transformation phase,* which is aimed at pursuing service innovation and e-government across a broader prism of community and democratic development through more networked governance patterns within government, across various government levels, and amongst all sectors in a particular jurisdiction. Governments must aim to reach the transformation phase, as this would provide more even ground for e-business and e-commerce growth (United Nations, 2008).

A different model of e-government development consists of five stages, namely emerging, enhanced, interactive, transactional, and connected. The *Emerging* stage is characterised by a government's online presence, mainly consisting of a web page and/or an official website; and links to ministries or departments of education, health, social welfare, labour and finance. Much of the information is static and there is little interaction with citizens. At the *Enhanced* stage, governments provide more information on public policy and governance. Here, they have links to archived information in a format that is easy for citizens to access and use, such as documents, forms, reports, laws and regulations, and newsletters. During the *Interactive* stage, governments deliver online services such as downloadable application forms for tax payments or license renewals. At this stage, the start of an interactive portal or website with services that enhance convenience for citizens are evident. The *Transactional* stage is characterised by governments' introduction of two-way interactions between citizens and government. It includes options for paying taxes, applying for ID cards, birth certificates, passports, license renewals, and other similar G2C interactions, and allows the citizen 24/7 access to these services online. All transactions are conducted online. Finally, the *Connected* (integrated or seamless) stage is characterised by a government's transformation into a connected entity that responds to the needs of its citizens by developing integrated back office infrastructure. This is the most sophisticated level of online e-government initiatives, and is characterized by (United Nations, 2008):

- Horizontal connections among government agencies
- Vertical connections between central and local government agencies
- Infrastructure connections with addressed interoperability issues
- Connections between governments and citizens
- Connections between various stakeholders, such as the government, private sector, academic institutions, NGOs and civil society
- The support and encouragement of e-participation and citizen engagement by governments in the decision-making process.

Governments can enhance the development of the digital economy by creating an environment in which the citizenry have more trust in their government. The interaction between a government and its citizens, and the responsiveness of the government to optimise its public service delivery outputs and outcomes within the constraints that it faces, have the potential to improve the trust with which [the government] is viewed by citizens. Building and maintaining trust in e-government requires an understanding of both the many levels of interactions where trust must be earned, and of cultural differences. Factors that affect trust relate to culture, information, content, and the nature of the system. It is therefore important to build citizen-centric e-government to uphold information rights that focus on human dignity, fundamental human rights, and earning trust. Building citizens' trust in e-government must take cognisance of ethical issues such as privacy, accuracy, and access. It is also important for leaders to promote knowledge sharing as a means of instilling trust among its employees and citizens (United Nations, 2008). Issues of privacy, confidentiality, data integrity and the availability of information are also important in environments where information is shared.

If privacy is not assured, people may not endear to e-government systems. Oak (2008) reminds us that Internet privacy is particularly important when website users give out their personal details on the Internet. For websites that facilitate online shopping, users are required to input their credit card numbers. Privacy in such an environment is of the utmost importance to instil trust in the perpetual use of the e-government services. Likewise, in the case of emailing sites where third parties are allowed to store or read emails without informed consent, there is likely to be distrust of the service among subscribers.

Citizens and businesses would also be encouraged to operate in an e-government environment if the information they seek and access is accurate and reliable. Information on government websites that contains errors will not meet the needs of citizens or businesses, who may feel that their interests and needs are not being taken seriously. Information on e-government portals may also link to information that does not necessarily belong to government; therefore governments should ensure that citizens and businesses accessing such information are insulated from any copyright violations that may arise thereof. In electronic commerce environments, businesses and customers expect to have quick access to information and direct responses to their requests. It is important for e-government systems to respond to this need.

Confidentiality ensures that only those with sufficient privileges and a demonstrated need may access certain information. For a business, exposing confidential information to theft could have a significant impact on matters such as stock prices, revenues and competitive advantages. Integrity, on the other hand, relates to the protection of content/information against corruption or the alteration of information. The integrity of information is threatened when it is exposed to corruption, damage, destruction or other disruptions to its authentic state. It is important for businesses to trust the mechanisms used for the storage and transmission of information in the electronic environment. E-government environments must ensure the integrity of their content at all times.

Since an e-government environment provides a platform for e-commerce transactions, a legal oriented framework is necessary to cater for cyber laws, consumer protection, and the security of transactions (Department of IT eTechnology Group -India, 2003). A consumer protection framework would deal with uncompetitive behaviour in the market and create a level playing field for all businesses. Investors in an e-economy would need assurance in the form of an environment that could enhance the profitability of their investments, and a competition policy is an important consideration in this respect. OECD countries have developed guidelines for consumer protection in the context of e-commerce that provide a solid base for consumers to receive the same level of protection when they shop online as they would if they bought from a local store (OECD, 1999).

Governments can cultivate trust in their citizens and businesses with regard to the use of e-government services when they take steps to cater for all in society, including the disabled and the elderly by, for example, implementing adaptive technologies. Governments also need to take into consideration their citizens' level of comfort with the various ICTs available in order to deliver effective online services. For the youth, this might mean providing online services via cellular phones and/or creating an efficient and robust portal that can respond to their need for speed and portability. For senior citizens, it might mean providing one-stop centres through which they can receive assistance on how to access online services without needing an even moderate knowledge of ICTs. For others, it might mean providing integrated portals whose back office operations are interlinked, thereby providing a seamless transition from one service to next (United Nations, 2008).

CONCLUSION

E-government and the digital economy are inextricably intertwined. Thus, in order for businesses to be able to effectively utilise e-government systems in the knowledge economy, an enabling environment is important. Southwood (2005) provides a list of questions that can assist in determining e-government (or e-business) readiness, as follows: Is the infrastructure for data systems ready? Are management systems, records and work processes in place? Is the legal infrastructure ready (e.g. the laws and regulations required permitting and supporting the move to e-government and facilitating the acceptance of digital signatures)? Are the institutional infrastructures needed to facilitate and drive e-government in place? Is our human infrastructure ready (i.e. do we have in place the attitudes, knowledge and skills required to initiate, implement and sustain e-government initiatives?)? Is our technological infrastructure ready? And is our leadership and strategic thinking ready (i.e. is there the internal authority and vision to drive forward e-governance?)?

Governments must take the necessary steps to address issues related to trust, especially privacy/confidentiality, security, integrity, availability, and universal access, in order to inspire businesses' use of electronic government systems. Moreover, a legal framework to ensure that electronic commercial transactions are enforced and electronic documents and signatures authenticated should be put in place. The protection of consumers in e-commerce contexts is important, as this provides a solid base from which to offer consumers online protection and security. National strategies that emphasise the interoperability of products and the affordable pricing of network and interactive content are good indicators of a government's commitment to an e-government system that can support business transactions effectively in an electronic environment.

REFERENCES

Adesharal, P., Juric, R., Kuljis, J., & Paul, R. (2004). A survey of acceptance of e-government services in the UK. *Journal of Computing and Information Technology–CIT, 12*(2), 143–150.

Bagire, V. (2005). *How e-governance breaks bureaucracy: Collaboration on international ICT policy for east and southern Africa (CIPESA)*. Nairobi: Kenya Government.

Bridges.org. (2001). *Comparison of e-readiness assessment models: Final draft*. Retrieved on July 16, 2003, from http://www.bridges.org/eredainess/tools.html

Chadwick, A., and C. May (2003). Interactions between states and citizens in the age of Internet: E-Government in the United States, Britain, and European Union. Governance, 16(2), 271–300. doi:10.1111/1468-0491.00216

Cloate, F. (2007, June 26-29). Knowledge management and trust in government: Lessons from South Africa. *Global Forum on Reinventing Government: Building Trust in Government*, Vienna, Austria.

Department of IT eTechnology Group (India). (2003). *Assessment of central ministries and departments: E-governance readiness assessment 2003: Draft report 48.*

Ganesan, A. (2008). *Global Internet freedom: Corporate responsibility and the rule of law: Testimony to the U.S. Senate Judiciary Committee, Subcommittee on Human Rights and the Law*. Retrieved on July 25, 2008, from http://hrw.org/english/docs/2008/05/20/usint18894.htm

Githongo, J. (1998). Tanzania's poor score on graft. Retrieved on July 18, 2005, from www.nationaudio.com/News/East Africa/2809/Regional/Regional.6.htm

Holzer, M., & Kim, S. T. (2005). Digital governance in municipalities worldwide: A longitudinal assessment of municipal Web sites across the world. Retrieved on February 15, 2007, from http://unpan1.un.org/intradoc/groups/public/documents/ASPA/UNPAN022839.pdf

New York State Archives. (2005). *Managing government records/e-government*. Retrieved on July 19, 2005, from http://www.archives.nysed.gov/a/nysaservices/ns_mgr_pub58_intro.shtml

Oak, M. (2008). Ethical issues of Internet privacy. Retrieved on July 26, 2008, from http://www.buzzle.com/articles/ethical-issues-of-internet-privacy.html

OECD. (1999). *The role of communications infrastructures in advancing electronic commerce*. Paris: OECD.

Onyancha, O. B. (2008, July 31-August 4). *E-Governance in Africa: Challenges, trends, and issues*. Paper Presented at Moi University 3rd Annual International Conference.

Oyomno, G. Z. (2003). *Towards a framework for assessing the maturity of government capabilities for e-government.* Johannesburg, South Africa: LINK Centre, Graduate School of Public and Development Management, University of the Witwatersrand.

Remmen. A. (2004). *Images of e-government–experiences from the digital North Denmark* Aaolborg: Department of Development and Planning, Aalborg University.

Rotberg, R. (2007). The Ibrahim index on African governance: How we achieved our rankings. African Business, 336, 20.

Sheridan, W., & Riley, T. B. (2006). *Comparing e-government and e-governance*. Retrieved on December 12, 2006, from http://www.electronicgov.net/pubs/research_papers/SheridanRileyComparEgov.d Southwood, R. (2005). E-Government special–does it exist in Africa and what can it do? *Balancing Act Issue, 93*.

UNDESA. (2006). *E-Government readiness assessment methodology*. Retrieved on December 12, 2006, from http://www.unpan.org/dpepa-kmb-eg-egovranda-ready.asp

United Nations. (2008). *UN e-government survey 2008: From e-government to connected government*. Retrieved on January 28, 2008, from http://unpan1.un.org/intradoc/groups/public/documents/UN/UNPAN028607.pdf

United Nations Divisions for Public Economics and Public Administration. (2002). *Benchmarking e-government: A global perspective*. New York: United Nations, DPEPA. Retrieved on July 31, 2002, from http://unpan1.un.org/intradoc/groups/public/documents/un/unpan003984.pdf

World Information Society Report. (2006). *Digital opportunity index 2005*. Retrieved on February 13, 2007, from http://www.itu.int/osg/spu/publications/worldinformationsociety/2006/World.pdf

Xiong, J. A. (2006). *Current status and needs of Chinese e-government users*. Carbondale, IL: Southern Illinois University.f

Chapter 10
E-Business and E-Commerce

INTRODUCTION

E-business and e-commerce are applications around which the digital economy revolves. The concept 'e-business' was originated by IBM in 1997 to refer to the sum total of buying and selling goods and services, servicing customers, and collaboration amongst business partners online. E-business consists of electronic applications that offer production planning, scheduling, outsourcing, and other business-related operation processes. E-commerce is still evolving; as of yet, a widely accepted definition has not been agreed on (Coppel, 2000). Loosely defined, e-commerce refers to doing business over the Internet; in other words the buying and selling of goods and services that can be delivered offline as well as the creation of products that can be digitised and delivered online (e.g. computer software). Activities in e-commerce may include order registration, electronic advertising, electronic billing, electronic marketing, online delivery and tracking, and customer services support.

DOI: 10.4018/978-1-60566-420-0.ch010

The concept of e-commerce appears to be more widely used than e-business, perhaps because more companies tend to transact electronically without indulging in customer support. Customer support may be contracted to a third party.

A practical example of the development of e-commerce can be illustrated in the form of a global engineering company that operates with a lean manufacturing process and has considerably reduced the travelling of its engineers to businesses/clients/suppliers to gather orders because this type of trip is now often completed through the Web. When an order is received, the tooling planner decides what tools are needed to process the order based on electronically transferred drawings. However, these drawings require the use of appropriate software so that the specifications of the drawings can be determined (Ehrlich, 2005). Before the web-based processing of orders was adopted, the fax machine was introduced and increased the span of communication, decreasing the time necessary for data transmission. However, it did not cater for prints; therefore the manual transfer of prints was still necessary. What followed was the development of Electronic Data Interchange (EDI). The introduction of EDI required the purchase of special technology in order to read the EDI files and interpret the data. EDI was an interface specially developed to electronically exchange data. Of course, this process has now been upgraded and replaced with web-based order processing. The engineering company, in this instance, can now utilize a web interface for each customer and retrieve their orders.

In its rudimentary form, e-commerce was initiated in the 1970s in the private sector before PCs were ushered into the public domain. The evolution of the Internet and the late 90s hype about e-government, coupled with pressure from businesses in anticipation of reaping great benefits, gave impetus to the growth of e-commerce, both in the private sector and in government. However, contrary to popular belief governments (or rather e-governments), are not the chief architects of e-commerce. Governments, motivated by the increased automation of their business transactions, are simply reengineering e-commerce business processes and adding them to their emerging e-applications, such as e-health, e-voting, online communities, e-procurement, e-taxation, e-judiciaries, etc. Both e-governments and e-commerce are public sector reforms that aim to enhance service delivery and good governance.

LeClaire (2005) divides the development of e-commerce into three stages. The 90s, at the height of the Internet revolution, is dubbed phase I. The period from 1996-2000 is known as the hype era or phase II, followed by the post-hype era or phase III between 2001 and 2004. Phase I was largely characterised by the growth of the Internet. At the time, much of the connectivity was in the form of slow dial-ups and online billboards. The most important development in Phase II was the emergence of the Netscape web browser in 1994, which opened the door for online retailers such as eBay and Amazon.com. Amazon was the first firm to offer online ordering services. Phase II of e-commerce was also characterised by increased online insecurity, particularly relating to lack of availability of service. This form of insecurity had a negative impact on consumer confidence. Thereafter, the Internet bubble burst, forcing many online service providers to close shop. The current phase of e-commerce, i.e. phase III, has been characterised by increasing confidence and trust in online shopping, thus once again enhancing online business growth. Increasingly, e-commerce is now witnessing the development of affordable, easy-to-use solutions that make it possible for entrepreneurs to transform their unique ideas into successful online businesses, such as online order fulfilment technologies and shipping.

While Electronic Data Interchange (EDI) - a standard for processing and transmitting information between computers over private value-added networks - existed for several years prior to the Internet and WWW commerce, its use and scope were (and still remain) limited. The slow growth of traditional e-commerce could therefore partly be attributed to the fact that e-commerce largely relied on such private, value-added networks, which were also characterised by high costs and limited connectivity because they operated on proprietary protocols and standards. Moreover, value-added networks relied on store and forward methods and not real-time transactions. E-commerce using EDI was suitable for the processing of batch transactions, such as passing purchase orders and invoices, but not for advertising and interactive applications because such applications require high bandwidths that EDI does not offer. On the plus side, value-added network connectivity provides relatively strong security, reliability and confirmation of receipt (Pyle, 1996).

TYPES OF E-COMMERCE

Various forms of e-commerce can be grouped according to the parties involved in the online purchase or sale of goods and services, i.e.: business-to-business (B2B), government-to-business (G2B), business-to-consumer (B2C), consumer-to-consumer (C2C), consumer-to-government (C2G), consumer-to-business (C2B), government-to-government (G2G), and public-to-government (P2G). Business-to-consumer (B2C), for example, refers to individual consumers purchasing products from manufactures or service providers; business-to-business (B2B) consists of sellers and buyers as business entities; and business-to-government (B2G) means that governments participate in e-commerce transactions with business entities, e.g. public procurement. Traditional types of e-commerce involve transactions between businesses and consumers (B2C) and between business partners (B2B).

E-commerce covers two main types of activities, namely indirect electronic commerce (electronic ordering of tangible goods; these must still be physically delivered using traditional channels such as postal services or commercial couriers), and direct electronic commerce (the online ordering, payment and delivery of intangible goods and services such as computer software, entertainment content, or information services on a global scale) [Fan, 2000]. The Internet's significant revolution and growth in the 90s has been largely responsible for the rapid growth in e-commerce sales, especially B2C. For example in the United States (where most e-commerce transactions take place), sales in the final quarter of 1999 were estimated at 0.7% of the country's total retail sales, while in Europe, B2C penetration in early 2000 was estimated at just over 0.2% of that region's retail sales (Coppel, 2000).

During the 90s, e-commerce implementation, spurred on by the growth of the Internet and the World Wide Web, engendered many opportunities and challenges that demanded prudent interventions at policy, legislative and regulatory levels. In 1991, the Internet had less than 3 million users around the world, and its application to e-commerce was non-existent. However by 1999, an estimated 250 million users accessed the Internet, and approximately one quarter of them made purchases online from electronic commerce sites that were worth approximately US$ 110 billion (Coppel, 2000). Come 2006, the number of Internet users globally was estimated at over 1 billion, while mobile phone users were estimated at 2 billion (Economist Intelligence Unit, 2006).

In a significant way, Internet commerce continues to change the way products are customised, distributed and exchanged, and how businesses and consumers search for and use products. However, e-commerce is still in its infancy and the opportunities that it engenders are yet to be fully exploited. In the United States and Europe, where e-commerce transactions are growing increasingly pervasive (especially in e-government environments), applications such as the online issuance of purchase orders; electronic signature of purchase orders; drivers license and I.D. card applications and renewals; vehicle registration; tax payment; and registration of changes in address; are becoming the norm rather than the exception (Boston Consulting Group, 1999).

Determinants of E-Commerce Adoption and Growth by and in, SMEs

There are several factors that determine whether or not SMEs can adopt e-commerce. Most SMEs will not adopt e-commerce if the benefits do not outweigh the costs of developing and maintaining the system. SMEs are generally concerned about costs because they generally suffer from budget constraints and are less sure of the expected returns on their investment. Some SMEs cannot afford to adopt sophisticated ICT solutions, for instance the purchase of secure web applications for credit card transactions. SMEs with a small number of employees may not need sophisticated websites, but might be encouraged to adopt a simple website without any e-commerce functionalities as long as they are able to afford the cost of basic Internet services (Bricklin, 2002). E-commerce maintenance and upgrades can be very costly, especially when firms aim for a highly sophisticated virtual shop on top of website maintenance, telephony, and Internet service provider (ISP) and site hosting charges. Moreover, website maintenance and upgrades may increase over time as the volume of online transactions grows, and this may raise costs and generate excessive workloads in terms of staff time. The price of Internet access is a key determinant in the level of Internet and e-commerce use by individuals and businesses. Countries with lower access costs typically have a greater number of Internet hosts, and electronic commerce has developed rapidly in countries with flat-rate access. OECD (2004) observes that for low-income and rural populations, universal access programmes are necessary to make adequate telecommunications available in areas where telecommunications operators cannot otherwise expect to provide network access.

A cross-country and cross-sector study by OECD's Electronic Commerce Business Impacts Project (EBIP) of 217 firms that were early adopters of the Internet and e-business strategies (OECD, 2002), revealed that these firms view competence factors such as management attitudes, skills levels and training, as the most likely to successfully influence businesses on the path to adopting e-business strategies. Technological factors, such as how to ensure interoperability with different e-commerce systems and improve network reliability and flexibility, were also highly favourable, although they were of less concern than competence factors. Cost factors, including the costs of reaching customers, engaging in e-commerce, and telecommunications, were also viewed positively in light of the benefits that can be gained from e-commerce and e-business. Comparatively, confidence factors such as brand image, transaction security, legal structures, and IPR issues, were of lower concern.

Policies that would affect the adoption and use of e-business strategies include those designed to expand and improve the quality of network infrastructure and the legal and regulatory environment, foster technological diffusion and create a favourable business environment. Specific policies for SMEs further tend to focus on ICT and e-business awareness programmes, business consultation services, and employee and management training to enhance ICT and managerial skills (OECD, 2004). ICT and e-commerce offer benefits for a wide range of business processes. For instance, ICTs enhance and speed up communication within and outside the firm. This increases the efficiency of the managers of the firms, especially if they are not in close proximity to the business. Sharing information electronically also enhances the efficiency of business processes such as ordering, invoicing and back-office operations. Through knowledge management systems, firms are able to store, share, and effectively use their acquired knowledge. The Internet allows SMEs to reach out to new customers, expand their markets geographically, and transmit and order information seamlessly between or across different systems. The Web also provides small players with an opportunity to join and compete in a wide variety of supply chains, including those previously inaccessible because of the use of costly closed EDI networks (Sakai, 2002).

The growth and development of e-commerce has been found to follow patterns similar to those of Internet adoption, assimilation and use across the world. Income, education and age are the main factors determining the profile of Internet users, especially B2C buyers. In the United States, surveys in 2003 showed that growth in the Internet population had occurred across every demographic group. However, there remained several factors that separated Internet users from lesser- or non-users. For example, younger Americans were significantly more wired than older Americans; well-to-do Americans were more wired than less well-off Americans; and the employed were far more wired than the unemployed. Moreover, White Americans were more wired than African-Americans and Hispanics; well-educated Americans were more wired than those who only completed high school; suburban and urban residents were more wired than rural residents; and the parents of children living at home were more wired than parents whose children were away (US Department of State, 2003).

The business use of the Internet also reveals a dichotomy, wherein usage is substantially higher among large than small-sized enterprises. For example in Japan, Internet penetration among firms with more than 300 employees was estimated to be as much as 80%, while in firms with less than six employees it was only 20% (Romer, 1990). Financial, legal and other service industries were found to have higher Internet penetration rates than those in manufacturing and mining. It has also been established that the relative level of capital spending on communications infrastructure and Internet applications' software development generally tends to be higher in countries with mature and liberalised telecommunications markets, for example the United States, United Kingdom, Japan, Finland, Sweden and Australia (Romer, 1990).

The increasing growth of electronic commerce over the Net has been motivated by factors that include, but are not limited to: pressure from business leaders across the world wishing to enhance their competitiveness and facilitate entry into global markets; global public sector reforms; the digital revolution, especially of the Internet and WWW; open nature of the Internet compared to traditional infrastructure such as value added networks; increasing use of ICT by government; the need for government to enforce tax compliance; and the increasing partnerships between different levels of government and the private sector. Payne (n.d) outlines some of the benefits of e-commerce to businesses, especially SMEs, as follows:

- It helps them find new customers, partners and suppliers, both domestically and internationally
- Serves current and new customers better, hence offering more value to them
- Improves the efficiency of the business' processes
- Offers entirely new services and products, some of which even lead to the start of new businesses.

Due to pressure from international monetary agencies such as the World Bank and the International Monetary Fund, governments are undertaking public sector reforms in order to reorganise the management of public procurement, and this process includes implementing e-government and e-commerce systems. Consequently, sizeable business-to-government (B2G) transactions are increasingly being implemented. Similarly, governments the world over are increasingly using the Internet for the transmission or receipt of information from business enterprises and citizens in order to improve service delivery and enforce tax compliance. Generally, there is also a growing increase in e-commerce worldwide, motivated by factors such as the low cost of using the Net. Through the Internet, buyers and sellers are drawn together in more efficient ways to create new market places and opportunities for the reorganisation of economic processes, which in essence is what e-commerce is all about.

Among the things that e-commerce is expected to do, are the following: stimulate productivity and economic growth; enhance efficiency in the management of the chains of supply and distribution; minimise barriers to entry into new markets; improve access to information that is customized to meet individual needs; and offer businesses, individuals or governments infrastructure that enables the creation of virtual markets for goods and services that hitherto never existed. Generally, e-commerce provides opportunities for doing business worldwide without the constraints of time and place.

SMEs have been found to use ICTs for e-commerce in applications such as communication via e-mail; providing information about a company's products, services and technologies via corporate websites; ordering and exchanges with regular customers; recruitment; and receiving customer feedback. Some SMEs also purchase standard materials, office equipment and software over the Internet.

While many studies provide evidence of the positive effects of ICT adoption on a firm's performance, others have shown no relationship between the two (Bitler,

2001). For example, although computers and access to the Internet have become common in most OECD countries, sales and purchases over the Internet have yet to take off. Furthermore, while the available data suggests that electronic commerce is growing, it still accounts for a relatively small portion of economic activity for firms of all sizes. Baldwin and Sabourin (2002) note that a study on Canadian manufacturing establishments with ten or more employees showed that those with higher growths in productivity were more likely to be using greater numbers of advanced ICTs. OECD analyses have also shown that the overall impacts of ICTs and e-business strategies on a firm's performance are positive; yet even then, ICTs are not a panacea in and of themselves. Most Internet e-commerce involves domestic rather than cross-border transactions. Baldwin and Sabourin (2002) suggest that most SMEs still appear to be at the point where establishing a website or adopting e-commerce is the main issue. The successful integration of external and internal business processes in e-business also necessitates organisational and managerial changes, which may entail proportionally greater costs and risks for SMEs. Furthermore, small-sized enterprises generally have fewer incentives to integrate their business processes than larger firms, which have more complex business processes and resources to harmonise and co-ordinate.

E-COMMERCE: E-GOVERNMENT NEXUS

Both e-commerce and e-government are public sector reforms that aim to improve service delivery to citizens and businesses and force governments to become more transparent and accountable in the way they carry out their businesses. Most work in government is of e-commerce nature because they entail the provision of services that largely involve monetary transactions, such as filing claims. Besides, both e-commerce and e-government enhance government-to-citizen and business-to-consumer service delivery by providing real-time access to information and transactions. For example, the one-stop e-government service concept provides citizens with single-faced government with which they can interact in seeking services. To meet the demand of one-stop government service, e-government systems usually maintain client data relating to the client's needs and situation. The benefits of e-commerce provision in e-government environment are that governments can eliminate middlemen and market their services directly to users by accepting orders via the Internet.

E-government and e-commerce application rely largely on ICT in the new global economy for the purpose of inter-linking governments to stakeholders such as businesses, communities locally and globally. This makes governments more competitive in both local and international markets. Moreover, businesses and governments also need to advertise and promote their services, and ICTs provide an important means by which this need can be achieved. ICT is the nexus-glue that links government and business, because it enables government (through e-government) and business community (through e-commerce) to do business together easily. The National Electronic Commerce Coordinating (2000) points out that in the United States, most transactions in government involve the exchange of services, products and money, making e-commerce a critical component of e-government. Moreover, the growth of e-commerce and the move towards e-government provides opportunities for governments and their citizens to interact with each other in a timely manner. The growing use of the Internet and the ability to buy goods and services anytime, whether day or night, has impacted on the way governments try to provide services to their citizens .and do business with the business community. Governments have attempted to ease the burden of a number of bureaucratic processes by allowing citizens and businesses to pay for and obtain certain services online. Citizens, in turn, are expecting more from their governments, and in certain circumstances, are willing to pay extra to avoid the inconvenience of doing business with government constituencies only during their business hours.

E-COMMERCE IMPLEMENTATION REQUIREMENTS IN AN E-GOVERNMENT SETTING

A healthy environment is necessary for the successful implementation of e-commerce. In particular (and especially in the telecommunications sector), economic and regulatory reforms that lower the cost and improve the quality of Internet access are crucial for e-commerce to thrive. Lower prices would, for example, stimulate greater computer diffusion into a country's households. Investment in network capacity and the speed of data transmission are also important for e-commerce uptake (Romer, 1990). In the business sense, when introducing e-commerce it is important to vigorously market the new solution across the entire organization and dedicate full-time project staff to the implementation process. Education and training are also key elements during the implementation process, and should not be forgotten.

When implementing e-commerce between government (through e-government) and the business community, it is important to address auditing and control issues. Auditors can assist to identify the key risks associated with these e-commerce applications, and also assess the effect of these risks on the audit work (The US' National Electronic Commerce Coordinating Council, 2000). As already mentioned, the design and operation of e-commerce systems in e-government need to address issues related to transparency and accountability. Moreover, the designers of e-commerce systems in an e-government environment need to understand that critical business-related issues should guide the transformation of traditional governments to e-government environments. Establishing effective IT infrastructure is the first step on the road to a well-controlled e-government environment. E-commerce implementations in e-government environments should also address leadership and governance; privacy and security; technology; legal and customer readiness; access; and applications and competencies (The US' National Electronic Commerce Coordinating Council, 2000).

E-commerce implementation in government is still a new innovation and is thus not yet that well understood. Consequently, strong leadership on the part of those charged with the responsibility of running governments is paramount. Also important in the e-commerce implementation process is the government's willingness to change traditional structures and business processes. These changes may be accomplished through legislations that reorganize the traditional government model or through strong IT leadership and policies. Governments need to be aware of potential cost-savings, and must hold their government agencies accountable for the achievement of the results enabled by new technologies. Moreover, successful e-commerce implementation requires adequate managerial support, adequate funding, sufficient project management, clearly defined project objectives, technology choices that meet business requirements, process integration, proper planning, aligning plans with practice, re-engineering business processes, ensuring that organizational structures lend themselves to coordination, and ensuring that organizational culture does not inhibit people's ability to perform. It is important to ensure that before e-commerce applications are implemented in a government's corridors, a written enterprise-wide strategic plan for e-commerce is promulgated and put in place. Such a plan would include a framework for the prioritization of business objectives and their related e-commerce activities; measurable results and milestones; the monitoring of activities; and a mechanism for identifying issues and initiating corrective action (National Electronic Commerce Coordinating Council- NECCC, 2000).

E-government, like the e-commerce environment, exacerbates issues of privacy. Given that governments collect large amounts of information on citizens and private enterprises, and given that this information is stored in electronic form, measures to enhance the privacy of these entities is crucial in order to instil confidence within parties interested in electronically transacting with governments. Some of the information that governments collect and store includes corporate filing information, tax information, and regulatory information. The implementation of e-commerce demands that privacy policies and laws are harmonized and made consistent; organizations dealing with government comply with privacy policies and laws; fraud scams, such as identity theft, do not occur; access to data is not inappropriately granted or refused; sensitive information is not disclosed without authorization and litigation; and legal liability does not proliferate. Overall, e-commerce implementation should ensure that consistent and legal privacy policies are established to overlook the e-government system. Moreover, governments should have a clearly written privacy policy defining the type of personal information collected and how it is collected; the intended use of and ultimate disposal of the collected data; and the identity of third parties, especially in cases where information is shared (National Electronic Commerce Coordinating Council, 2000).

As with privacy, security issues cause great concern for the parties involved in e-commerce transactions. With more citizens and businesses submitting information to governments over the Net, the risk of aspects of it being stolen increases. The type of information most liable to theft includes user identification, passwords, credit card numbers, bank account numbers, etc. In the event of the loss or theft of such information, governments could be liable to litigation. It is therefore important to have in place an effective security system that protects data when it is stored and during transmission to or from government.

E-commerce within governments involves various parties, such as citizens, private sectors, other governments, suppliers, and international players with different systems and laws governing their operations. Successful e-commerce implementation in governments therefore demands interconnectedness and the standardization of hardware and software to ensure compatible communications across different systems. Systems should be developed with enough flexibility to accommodate changes in technology and provide a sustained ability to communicate. E-commerce's technology implementations should ensure that systems interface with each other appropriately to enhance efficient processes. Moreover, systems need to be secured and unproven technologies should not be used, as these could lead to system failure. The systems deployed should also not have a single point of failure that would interfere with a system's operations or business continuity in the event of a catastrophe. In order to achieve this, security requirements demand, among other things, establishing methods that ensure uninterrupted service delivery, such as maintaining a mirror image, redundant communication lines, and other related backup facilities (National Electronic Commerce Coordinating Council, 2000).

In order for e-commerce implementation to have the desired impact on government, there is a need to ensure that the various stakeholders with whom the government is dealing are ready for e-commerce operations. A major problem has always been that all sectors of society do not benefit equally from ICTs because not all sectors of society have equal access to technology. The so-called digital divide manifests itself in various forms, such as the inability to afford or access a computer; disabilities, e.g. visual impairment; linguistic problems; lack of Internet access for those living in rural areas or inner city neighbourhoods; computer illiteracy; and no access to bank accounts or credit cards for prolonged or online payment. The implementation of e-commerce further requires adequate and well equipped human resources. The competency of the individuals developing, implementing, and supporting technology drives the effectiveness and efficiency that can be achieved through e-commerce and e-government. Moreover, personnel should be in possession of proper skills in order to perform as expected.

Alliance for Global Business (1999) provides a comprehensive checklist of 'dos' if a digital economy that is driven by e-commerce is to succeed. It points out that the development of a digital economy is more likely to be successful if the development of electronic commerce is led primarily by the private sector in response to market forces. Moreover, participation in electronic commerce needs to be pursued through an open and fair competitive market. Government intervention, when required, should support a stable, international and legal environment, promote the fair allocation of scarce resources, and protect public interest. Such an intervention needs to be clear, transparent, objective, non-discriminatory, proportional, flexible, and technologically neutral. Mechanisms for private sector input and involvement in policy-making should be promoted and widely used in all countries and international forums. Because electronic commerce is global, the government policies that affect it should be internationally coordinated and compatible. Such policies should also facilitate interoperability.

The transactions conducted using e-commerce should receive neutral tax treatment in comparison to non-electronic transactions. The taxation of electronic commerce has to be consistent with established and internationally accepted practices, and administered in the least burdensome manner. Regulation of the underlying telecommunications infrastructure, when necessary, could enable actors to compete globally in an open and fair market. The protection of users, in particular with regard to privacy, confidentiality, anonymity and content control, should also be pursued through appropriate policies. A high level of trust in the global information infrastructure can be encouraged through mutual agreement, education, furthering technological innovations that enhance security and reliability, the adoption of adequate dispute resolution mechanisms, and private sector self-regulation (Alliance for Global Business, 1999).

E-Commerce Readiness

The subject of e-readiness has already been addressed in chapter six, but will be repeated in this section in the context of e-commerce readiness. Likewise, some of the e-readiness tools that have already been covered will be repeated here in the context of assessing e-commerce readiness. The concept of e-commerce readiness is hardly defined explicitly in literature. The Computer Systems Policy Project (CSPP) refers to it as the availability of efficient telecommunications services and IT industries; the supporting environment of education, business and government; and a relevant policy and legal environment (Pentland and Figueres, 1999). The Economist Intelligence Unit/Pyramid Research (2000) describes e-readiness as the extent to which a country's business environment is conducive to Internet-based commercial opportunities. It is a concept that spans a wide range of factors, from the sophistication of the telecoms infrastructure, to the security of credit-card transactions and the literacy of the population. The Economist Intelligence Unit/Pyramid Research categorizes countries according to their level of readiness, beginning with e-business leaders, which are countries that already have most of the elements of e-readiness in place, although there are still some concerns about regulatory measures. E-business contenders are countries that have both satisfactory infrastructure and a good business environment. However, parts of the e-business equation in this category are still lacking. E-business followers constitute the largest group of countries, and while they have started to create an environment conducive to e-business, they still have a great deal of work ahead. Finally, the business laggards are countries that risk being left behind and face major obstacles to e-business growth, primarily in the area of connectivity.

In determining the level of e-readiness, the Economist Intelligence Unit/Pyramid Research (2000) weighs factors that they believe determine whether or not a country is prepared to seize the opportunities presented by the Net when guided by an overall positive business climate. The use of specific elements of Internet and e-business infrastructure, such as connectivity, social and cultural factors, the legal environment for e-business, the business environment, the development of e-commerce, and the existence of supporting e-services, form six categories that feed into their rankings. Specifically, the categories are as follows:

Connectivity (30%): They point out that e-business thrives on adequate telecommunications and Internet infrastructure; consequently they use connectivity measures to assess the access that individuals and businesses have to basic fixed and mobile telephony services, including voice, narrowband and broadband data. Affordability and the availability of services are also important determinants of connectivity.

Business Environment (20%): This aspect covers the general business climate based on 70 indicators, including criteria such as the strength of the economy, political stability, the regulatory environment, taxation, and openness to trade and investment.

E-commerce Consumer and Business Adoption (20%): This measure consists of payment and logistics systems that evaluate the extent of credit-card ownership; the existence of secure, reliable and efficient electronic payment mechanisms; the ability of vendors to ensure the timely and reliable delivery of goods; and the extent to which websites have been developed by local firms.

Legal and Regulatory Environment (15%): The legal framework governing e-business is considered to be a vital factor that can either enhance or inhibit the development of electronic trading. Consideration here is given to the extent of legal support for virtual transactions and digital signatures, the ease of licensing, and the ability of firms to operate with a minimal but effective degree of regulation.

Supporting E-services (10%): EIU/Pyramid Research point out that no business or industry can function efficiently without intermediaries and ancillary services for support. For e-business markets, these include portals and other online intermediaries, web-hosting firms, application service providers, web developers and e-business consultants.

Social and Cultural Infrastructure (5%): This aspect considers education and literacy to be necessary preconditions for a population's ability to navigate the web and drive future domestic Internet development. Here, the nation's proclivity to business innovation and receptiveness to web content are assessed.

Doyle et al. (1999) outline how national states can be empowered to take part in the digital economy as follows: educating government officials, top civil servants, and voters on the nature of the electronic economy; cooperating internationally in order to regulate consumer protection and develop a priority list of network crimes and agreed punishments; investing or encouraging investment in the physical infrastructure of the electronic economy, such as interactive networks and terminal access equipment; equipping citizens with the means to log on to the global network of wealth, information, and power, recognizing that there is a direct relationship between network access and the wealth of nations; replacing the measures and statistics of the industrial economy with the measures and statistics of the electronic economy, especially seeing as the official statistics, which form the basis of trade negotiations, are derived from an outdated data collection model; and providing individual consumers and citizens access to information networks as a right rather than a privilege. Other measures to enhance e-readiness in the digital economy include (Doyle et al. 1999):

- Appointing a chief public information officer to act as a champion of the public information used to support consumers' and citizens' rights
- Using electronic tools to support government processes, such as social provision, education services, and public procurement
- Developing inward investment programs to attract and develop knowledge industries based on electronic networks
- Making connectivity, network infrastructure, taxation regimes, the price of bandwidth, education and information services attractive to these enterprises
- Intervening directly to guarantee and regulate the digital exchange of money.

World Information Technology Services and Alliance (2000) summarizes the main variables to consider in assessing the preparedness of a nation to partake in the digital economy as the affordability or reliability of network access, including the cost of services, downtime, and the prevalence of access among individuals. In addition, the reliability of the electrical supply for business-critical computer operations, and the ease of importing, exporting and transporting goods within or outside a country, are important considerations. Still others include (World Information Technology Services and Alliance, 2000):

Leadership: The priority given by government to promoting the development of an e-society on a national level; the extent of demonstrated progress on e-government, including efforts to automate governmental processes, offer services to businesses and citizens electronically, and create national portals; quality of partnerships between industry leaders and government to improve e-readiness; and level of effort put into promoting access for all citizens.

Information Security: Strength of legal protection and progress in protecting intellectual property rights, especially of software; extent of efforts to protect electronic privacy; and strength and effectiveness of the legal framework to address and prosecute computer crimes, authorize digital signatures, and enable public key infrastructure.

Human Capital: Quality of and participation levels in the education system, with an emphasis on efforts to create and support a knowledge-based society; penetration of ICT in schools and ability of educators to use and teach in accordance with the technologies; culture of local creativity and information sharing within the society; skills and efficiency of the workforce; and strength of efforts to retain skilled managers and technologies.

E-Business Climate: The existence of effective competition among communication and information service providers; transparency and predictability of regulatory implementation, openness of government, rule of the law, and general business risks (e.g. political stability, financial soundness); openness to financial and personal participation by foreign investors in ICT businesses; ability of the financial system to support electronic transactions; and sponsorship of science and technology facilities or parks as hubs of innovation and support for new enterprises.

ILO (2001) states that countries with the right mix of skills stand a better chance of becoming primary locations in global markets. Thus in order for maximum gains to emerge, the development of essential ICT skills is necessary, because without such skills, the technologies can neither be maintained nor adapted to local use. The adoption of ICT in business environments creates two types of skills needs. The first is related to a variety of foundation skills, such as the ability to communicate and analyse and solve problems. The other skills relate to the technical component, which extends beyond the ICT sector to the economy at large. The importance of developing ICT skills has been underscored by Zakaria (2001), who observes that the widespread use of emerging technologies, both at work and at home, demands flexible workers who are able to keep up with technology, and who are self-directed and knowledgeable about the world.

There is also a need to create and enhance awareness about the importance of using ICT for business competitiveness. Among developed and developing countries, there are wide variations in the level of awareness and skills' capacity in the application of ICT for economic development. For example, while a study carried out in Canada found that several business firms were reportedly aware of the considerable benefits of e-commerce, such as gaining new customers and reducing transaction costs (Agriculture and Food Canada, 2001), a study carried out in South Africa in 2003 to determine the effect of IT on the productivity of small-sized enterprises revealed, among other things, that very little was known about the technological needs of the business sector (Gordon, 2003).

METHODOLOGIES FOR ASSESSING E-COMMERCE READINESS

The subject of e-readiness in general has already been covered in this book, but the purpose of its exposition here is simply to show how e-commerce, as a specific application of the digital economy, can be surmised and measured. The concept of e-readiness, as already pointed out, refers to the extent to which a market is conducive to Internet-based opportunities. It focuses on the ability of consumers, businesses and governments to use ICTs to their benefit. Indicators for e-readiness include the quality of IT infrastructure, the degree to which the

Internet is creating real commercial efficiencies, and the policy and legislative frameworks in place.

Assessing e-commerce readiness presents several obstacles, first because there is hardly a standardised definition of e-commerce. Moreover, there are as yet no well established standard methods for measuring the level and growth of e-commerce, or for that matter, standardised methods for comparing statistics. It is therefore difficult to measure the impact of e-commerce on variables such as productivity, employment, new skills, etc (Colecchia and Schreyer, 2002). Because of the lack of well established models for assessment, a combination of approaches is generally used to surmise the level of e-commerce readiness. These approaches comprise of one or more of the following: digital opportunity index, e-readiness assessments, e-government maturity levels, e-government indices, and the information society index.

OECD member states use the following indicators to assess levels of e-commerce: number of ISPs; number of telecommunications carriers; telephone and digital fixed access lines; Internet, extranet, and intranet access; perceived obstacles to e-commerce; perceived benefits of e-commerce; number of persons with computer skills; expenditure on e-commerce; Internet access baskets; Internet hosts; secure servers; e-commerce readiness - technical and commercial infrastructures; e-commerce intensity - usage, volume, value and nature of transactions; identifying who is taking advantage of e-commerce opportunities and who is not.; and e-commerce impact - efficiency and/or wealth creation (OECD, 1999).

Likewise, the World Summit on Information Society's (WSIS) Declaration of Principles provides good prescriptions for determining e-commerce readiness. These indicators (WSIS Civil Society Working Group, 2004) include: secure networks; protection of privacy and confidentiality; promoting the public's awareness of ICTs; education and training programmes in ICTs; bridging the digital divide; universal access; open standards; predictable policies; legal and regulatory framework(s); and the availability of appropriate electrical power sources. The eEurope Action Plan (2004), on the other hand, uses the following indicators to assess e-commerce readiness: access and effective use of ICT; ICT skills development; promotion of e-commerce in both the public and private sectors; reshaping the regulatory environment for communication networks and services; opening doors to new generations of mobile and multimedia services; and having in place a legal framework that caters for e-commerce in the member states.

The Economist Intelligence Unit (2006) carries out an annual e-readiness survey of the world's major economies. In 2006, for example, 68 countries were surveyed, and the results revealed that most developed economies are well placed to compete in the global digital economy. Countries from these economies include Denmark,

The United States, Switzerland, Sweden, the United Kingdom, the Netherlands, Finland, Australia, Canada and Hong Kong. Overall, Europe remains the most dominant region in the world. Most of these countries are taking some form of action to improve their countries' e-business environment, by, for example, investing significantly in broadband wireless technologies such as WiFi and WiMax to improve online access. Mobile Internet access has also gained favour as customers in these countries increasingly become comfortable with mobile devices. Some countries have put Voice over IP (VoIP) in place to enhance connectivity, as VoIP reduces the cost of international calls for consumers and businesses alike.

E-Commerce Challenges

Small-sized enterprises would be expected to face a number of challenges in the e-commerce environment. As already pointed out, beyond a certain level of connectivity (PC, Internet access, online information or marketing), not all SMEs will necessarily compete against large firms, quite simply because e-commerce may not bring about large benefits; hence these SMEs choose to stay within the confines of traditional business processes. Moreover, the availability of ICT competencies within the firm, and the cost of appropriate interoperable systems, network infrastructure, and Internet-related support services, are barriers to most SMEs' entry into the global e-commerce market. Lack of reliable and trustworthy redress systems, and cross-country legal and regulatory differences, also impede cross-border transactions (OECD, 2004).

Several studies have been undertaken on the challenges facing SMEs in the digital economy. CommerceNet (2000), in a survey involving over 1000 respondents from six countries in North America and Asia, found the following to be the top barriers to electronic commerce: security and encryption; trust and risk; lack of qualified personnel; lack of business models; user authentication and lack of key public infrastructure; fraud and risk of loss; slow Internet - unreliability of the Internet; legal issues; interoperability issues with regard to legacy systems; high cost of entry; inconsistent tax laws; proprietary technology; and lack of information resources and decision support tools tailored to meet the needs of SMEs. In another survey by the World Information Technology and Services Alliance (n.d) of 27 countries worldwide, the following issues in a digital economy with regard to e-commerce were identified:

Trust: The survey found that developing countries cited low levels of credit card use and restrictions on using credit cards over the telephone as barriers to implementing consumer electronic commerce. Privacy ranked next among their concerns, followed by authentication – being sure of the identity and credentials of the party you are communicating with.

Technology: Respondents identified a wide range of technological barriers, including the poor quality of security systems; the efforts needed to integrate electronic commerce systems with existing enterprise systems; the lack of internationally recognized standards covering activities such as transaction processing, security and authentication; and limited networking bandwidth.

Workforce Issues: The survey also established that a shortage of skilled workers in the IT industry was the most important issue for most of the respondents. With regard to the leadership of business executives, it was expressed that those who run enterprises are often averse to taking risks.

Public Policy: Leading public policy issues, as highlighted by the respondents, were the following: the development of standards for authentication that would ensure trading partners are legitimate; the impact on electronic commerce of the taxation of online sales; and the confusion caused by conflicting international contractual and legal frameworks. Respondents also pointed to limits on the use of encryption by governments concerned with national security and fighting crime.

Taxation: The survey established that local sales taxes were perceived by respondents as most harmful to electronic commerce.

Business Processes: The survey found that the respondents expressed fear of opening corporate systems to outsiders, whether they were customers or suppliers. In addition, there was a lack of business models for newcomers to adopt; the logistical challenges of the real time environment of electronic commerce; and the need to be sure of quality business results from electronic information exchanges.

Costs: The costs of changing business processes and adapting corporate cultures in order to accommodate electronic commerce were seen to be a major obstacle.

Consumer Attitudes: The survey established that there was fear of committing personal information, such as credit card numbers, addresses and telephone numbers, to cyberspace, or losing money by purchasing goods from unknown companies; and the absence of regulations to govern procedures in the event of disputes were also cited as reasons for being wary of electronic commerce.

The virtual environment of electronic markets makes it more difficult to determine who the contracting parties are, where an electronic commerce operator is located, and whether that operator is complying with the relevant legal obligations and regulatory regimes. This can create legal and regulatory uncertainty. For example, regulations on advertising outlaw the use of English in France, advertising to children in Denmark, and comparative advertising in Germany; yet the content on the Internet is essentially borderless. It is not yet clear how these countries intend to police such rules on the Internet (Coppel, 2000).

The absence of commercial codes and legal recognition covering areas such as the acceptance of electronic signatures and documents; contract enforcement; and greater certainty vis-à-vis liability for damages that may arise as a result of electronic transactions; limits the take-up of e-commerce, especially when trading across borders. There are also issues of Internet governance, because an increase in electronic commerce would expand the number of Internet addresses required and accelerate the need for further reforms of the Domain Name System (DNS). Such addresses should be easy to acquire, with mechanisms for dispute resolution and provisions for expansion. The price dispersion across Internet markets and the absence of noticeable price reductions have led to concerns that the cost structure of some Internet markets could ultimately result in less competitive outcomes, especially among digital and knowledge intensive products. Once the first copy of a software application, for example, is produced; the cost of a second copy is close to zero. Such a cost structure implies increasing economies of scale. The challenge to firms is to find a way to price their output such that they may sell to a broad enough audience and thereby recoup the high initial per unit cost of production. Part of the reasons for this can be attributed to the substantial costs to merchants of setting up the necessary facilities. There also appears to be a lack of acceptance on the part of the public stemming from security and privacy concerns, since most systems can keep track of what users buy (Coppel, 2000).

There is a growing need for open standards in terms of the formats used to present text, images, video and other collections of data, so that one producer's data can be accessed using another's software. In addition to standards for the distribution and manipulation of information, standards are necessary for networked economic transactions - the actual exchange of money for digital goods. Varian (1996) raises a couple of questions regarding the difficulties associated with the lack of e-commerce enabled infrastructure and the difficulty of pricing digital commodities, e.g.: Will users subscribe to information services, or will they be able to buy data spontaneously? And how will payment be divided among the various parties involved in the transaction, such as authors, publishers, libraries, online services and so on? Moreover, if intellectual property protection is too lax, there may be inadequate incentives to produce new electronic works, and if the protection is too strict, it may impede on the free flow and fair use of information. Therefore, copyright and patent laws that apply to digital technologies need to be carefully thought out before being put in place. The shared nature of information technology makes it critical to address issues of standardization and interoperability.

The digital divide is a hindrance to sustainable development in any economy in the new millennium, and especially to SMEs. There is a need for businesses to change their ways and react quickly and effectively to benefit from the digital economy, because the demands of the digital economy are different from what businesses have been used to in the bygone agrarian and industrial eras. These demands include trusting the vendors of goods or services; dealing with the impersonality of transactions; having back up facilities for the processing of electronically received orders or electronic payments; providing after sales services in remote locations; dealing with domain name infringements, violation of trademarks and patents, and copyright laws; and settling revenue sharing disputes on royalty. To ensure SMEs' effective participation in e-commerce, there is a need for mentorship in the start-up phase and access to sustainable finance/credit at reasonable rates, particularly in critical times. SMEs also face problems related to limited access to state-of-the-art technology; dynamic avenues in marketing; unequal competition; inadequate infrastructure support; limited knowledge of the market; stringent government or bank procedures; lack of managerial skills; limited human resources; and biased legislation in favour of large enterprises, among others (Tuteja, 2000).

CONCLUSION

E-commerce transactions revolving around B2B, B2C, B2G, C2B, C2C and P2G form the largest proportion of the digital economy's business activities. The rapidly growing e-commerce industry, as a major component of the digital economy, has been given impetus by the open nature of the Internet in terms of limited regulation and the low costs involved, especially when compared to the more expensive form of EDI e-commerce that characterised the 1970s and 1980s. A number of factors are determining e-commerce adoption in the digital economy, such as education and income levels; age and size of the enterprises (with large enterprises more likely to adopt e-commerce than small-sized enterprises); security of transacting business online; and levels of e-readiness. Various challenges constitute barriers to e-commerce growth and development, including issues of trust, policy and regulation, privacy, skills, attitudes, etc. The challenges preventing SMEs from effectively participating in e-commerce require government interventions and policy and regulatory frameworks that would create an environment conducive to taking part in the global digital economy.

In terms of policy and managerial implications, governments can take several steps to promote e-commerce. Governments should first provide leadership by engaging in e-commerce. Because they tend to be the largest buyers of goods and services in many countries, their participation in e-commerce can immensely enhance its (e-commerce's) growth. Moreover, if governments were to buy a great deal of their goods and services from domestic SMEs, they would enhance the growth of the industry and inspire other consumers to do likewise. Governments can also enhance e-commerce by, for example, using e-procurement techniques, and in so doing provide incentives for SMEs to do likewise. By enhancing e-governance, governments can get businesses to turn to them for a wide variety of services, such as customs clearances, business licenses, dispute resolution mechanisms, etc. Governments must also put in place a comprehensive policy and legislative strategy that addresses, in an integrated manner, the various policy, legal, market, technological, and social considerations that interact at both domestic and global levels to create conditions conducive for e-commerce-led growth. Legislation and policies that cover issues on e-commerce, cyber crime and security, privacy and data protection, competition, intellectual property rights, consumer protection, the digital divide and e-governance are necessary.

UNDP (2007) identifies the kind of policy that would enhance SMEs' participation in the global digital economy. As a basic requirement, a healthy business environment is fundamental for firms to thrive and benefit from the digital economy. Such an environment would require: a transparent, open, and competitive business framework; a clear and independent rule of law, covering all firms; a way to easily set up and dissolve businesses; transparent, simple and accessible corporate regulations; and equal and stable legal treatment for national and cross-border transactions. Governments must develop and implement policies and regulations that would remove barriers to help businesses become more profitable and competitive. A key component in ICT development, adoption and use is broadband connectivity. This factor accelerates the contribution of ICTs to economic growth, facilitates innovation, and promotes efficiency, network effects and positive externalities. The effective deployment and use of broadband services require policies that: promote competition and continue to stress the liberalization of infrastructure; network services and applications across different technological platforms; encourage investment in new technological infrastructure, content and applications; and take a technology-neutral stance among competing and developing technologies to encourage interoperability and innovation, and to expand choice.

A regulatory trust is important for SMEs to address security, privacy and consumer protection interests. Issues that need to be considered here include: cyber-crimes and cyber-security; cross-border cooperation; and the presence of low-cost online dispute resolution mechanisms. This is followed by human capital development and skills enhancement. Lack of ICT and business skills are widespread impediments to effective uptake once decisions to adopt e-commerce are made. Governments have a major role to play in introducing basic ICT skills to primary and secondary schooling, and a further important role in conjunction with education institutions, businesses and individuals in providing the framework to encourage ICT skill formulation at higher levels, in vocational training, and in ongoing, lifelong learning. Moreover, the online provision of government information and services can increase the efficiency and coverage of public service delivery to SMEs.

Furthermore, as model- and standard-setting users of broadband, governments can convincingly demonstrate the potential of broadband-based services and content as well as the benefits of ICT adoption to small-sized enterprises, thus helping spread new services more widely. Education, general government information and services, and the provision of specific government services to businesses and citizens, can all benefit from the use of new high-speed infrastructure and services, and this should therefore be given priority in government strategies. Finally, public-private partnerships are paramount to developing effective e-business policies for SMEs. In public-private partnerships, governments' primary role is to provide an enabling policy and legal and regulatory environment that levels the playing field and allows the private sector to compete freely and fairly with everyone else (UNDP, 2007).

REFERENCES

Agriculture and Food Canada. (2001). *The e-readiness of Canadian Agriculture and Food Canada SMEs*. Retrieved on March 21, 2004, from http://www.agr.gc.ca/misb/fb/food/ecommerce/eready/introduction_e.html

Alliance for Global Business. (1999, October). *A global action plan for electronic commerce. Prepared by business with recommendations for governments, 2nd ed.*

Baldwin, J. R., & Sabourin, D. (2002). *Impact of the adoption of advanced ICTs on firm performance in the Canadian manufacturing sector*. Paris: OECD.

Bitler, M. P. (2001). *Small businesses and computers: Adoption and performance: Preliminary draft*.

Boston Consulting Group. (1999). *The race for online riches--e-retailing in Europe*. Maryland: Silver Spring.

Bricklin, D. (2002). Small business and Web sites. Retrieved on October 12, 2008, from http://www.bricklin.com/smallbusiness.htm

Colecchia, A., & Schreyer, P. (2002). The contribution of information and communication technologies to economic growth in nine OECD countries. *OECD Economic Studies,* (34), 153-170.

CommerceNet. (2000). *Barriers to electronic commerce 2000 study*. Retrieved on March 14, 2007, from http://www.oema.org/EC-Best-Practice/Research/Electronic Commerce Research/Other Sources.htm

Coppel, J. (2000). *E-commerce: Impacts and policy challenges*. Economics department working paper no.252. Retrieved on July 20, 2007, from http://www.oecd.org/eco/eco

Doyle, C. J., Ferguson, G. T., & Morris, H. F. (1999). *Ten ways national states can participate in the electronic economy*. Retrieved July 19, 2008, from http://www.ifc.org/publications/pubs/impwt99/effect/w9effect.html.

Economist Intelligence Unit. (2006). Digital divide narrows. *The Economist*. Retrieved on February 18, 2007, from http://www.ibm.com/news/be/en/2006/04/2601.html

Ehrlich, D. M. (2005, August 11-14). A study concerning small manufacturing business within B2B e-commerce value chain. In *Proceedings of the Eleventh Americas Conference on Information Systems*, Omaha, NE.

Fan, Q. (2000). *On the way to e-commerce: Chinese experiences (1998-1999)*. Retrieved on February 23, 2007, from http://ausweb.scu.edu.au/aw2k/papers/fan/paper.html

Gordon, G. (2003, February 16). SME survey. *Sunday Times Business Times*, 15.

ILO. (2001). Life at work in the information economy. *World Employment Report 2001*. Retrieved on March 11, 2004, from http://www.ilo.org/public/english/support/publ/wer/overview.htm

LeClaire, J. (2005). The evolution of e-commerce. *E-commerce Times*. Retrieved on February 21, 2006, from http://www.ecommercetimes.com/story/40249.html

National Electronic Commerce Coordinating Council (2000). A risk assessment guidebook for e-commerce/egovernment. Retrieved on February 2, 2007, from http://www.ec3.org/Downloads/2000/Risk_Assessment_Guidebook.pdf

OECD. (1999). *The economic and social impact of electronic commerce--preliminary findings and research agenda*. Paris: OECD.

OECD. (2002). *The impacts of electronic commerce on business: Summary DSTI/ICCP/IE (2002)5/FINAL*. Paris: OECD.

OECD. (2004). *ICT, e-business, and SMES*. Paris: Publications Service.

Payne, J. E. *E-commerce readiness for SMEs in developing countries: A guide for development professionals*. Washington, D.C.: Academy for Educational Development.

Pentland, S., & Figueres, J. M. (1999). *E-commerce notes*. Retrieved on February 23, 2007, from http://alumni.media.mit.edu/~mikeb/dev-ec/ecommerce.htm

Pyle, R. (1996). Electronic commerce and the Internet. Communications of the ACM, 39(6), 22-23. doi:10.1145/228503.228507

Ramsay, E., P. Ibbotson, J. Bell, and B. Gray (2003). E-Opportunities of service sector SMEs: An Irish cross-border study. Journal of Small Business and Enterprise Development, 10(3), 250–264. doi:10.1108/14626000310489709

Romer, P. M. (1990). Endogenous technical change. *Journal of Political Econom*, 71-103.

Sakai, K. (2002). *Global industrial restructuring: Implications for small firms. STI working papers* 2002/4. Paris: OECD. Retrieved on October 12, 2008, from http://www.oecd.org/sti/working-papers

The Economist Intelligence Unit/Pyramid Research. (2000). *E-readiness rankings*. London: EIU.

Tuteja, S. K. (2000, September 28-29). *Small business survival & growth in the context of globalization. The Indian scenario*. Paper presented at the APO Study Meeting on Small Business Survival & Growth, Jakarta, Indonesia.

UNDP. (2007). *The role of governments in promoting ICT access and use by SMEs*. Retrieved on November 18, 2008, from http://www.apdip.net/apdipenote/12.pdf/

U.S. Department of State. (2003). *Survey shows ever-changing patterns in U.S. Internet use*. Retrieved on February 23, 2007, from http://www.iwar.org.uk/news-archive/2003/04-18.htm

Varian, H. R. (1996). The information economy: How much will two bits be worth in the digital marketplace? *Educom Review*, 44-46. Retrieved on March 18, 2007, from http://www.educom.edu

World Information Technology and Services Alliance. (2000). The world information technology and services alliance international survey of electronic commerce. Retrieved on April 12, 2007, from http://www.witsa.org

WSIS Civil Society Working Group. (2004). *World summit on information society: Draft declaration of principles*. Retrieved on November 7, 2004, from http://www.wsis-pct.org/prepcom2-declaration.html

Zakaria, Z. (2001). *Factors related to information technology implementation in the Malaysian Ministry of Education Polytechnics*. Unpublished doctoral dissertation, Faculty of the Virginia Polytechnic Institute and State University, Blacksburgh, VA.

Section 5
Content, Knowledge, and Records Management

This section consists of three chapters: Chapters 11 (Local Content and SMEs), Chapter 12 (Knowledge Management), and Chapter 13 (E-Records Management and Freedom of Information for SMEs). Chapter 11 begins by unpacking the concept of content and its derivative, local content. The chapter then discusses the value of local content in enhancing business competitiveness. Also discussed are local content development initiatives at international, regional, and national levels, with mention of the agencies that are involved in these initiatives. Challenges of local content development are explained, as are the prospects of local content for SMEs. The chapter also elucidates on policy and managerial roles in local content management and its applications to SMEs. Overall, the chapter aims to demonstrate the value of relying on locally available products and services, in part because they would be cheaper and more easily understood.

Chapter 12 reviews the growing number of studies on knowledge management (KM) as they apply to SMEs. The chapter covers the benefits SMEs can accrue from knowledge management, the meaning and evolution of KM, KM strategies, KM general adoption practices, KM processes and activities, KM technologies, KM challenges, and the policy and managerial responsibilities of KM. The chapter aims to promote awareness among SMEs, governments and SMEs' managers of the value of harnessing knowledge, human capital and know-how for competitive advantage in increasingly competitive electronic business environments. Given that the digital economy is knowledge intensive, SMEs must wake to the fact that without investing in human capital and effectively managing the intellectual capital already in their possession, they may not be able to compete in the digital economy.

Chapter 13 introduces the concept of e-records management and observes that electronic records management has not been given much attention by SMEs, this despite an economy where most transactions occur online, meaning that e-records are increasingly being generated and must be meticulously managed. The concept of freedom of information (FOI) is discussed in the context of how it facilitates access to publicly held information, which is necessary for promoting business activities within SMEs. The chapter also covers the role of e-records management, the nexus between records management, SMEs, and FOI, global initiatives for enhancing FOI, e-records readiness, e-records generation and use, e-records management, e-records management models, e-records and information security, e-records authentication-digital signatures, digital certificates, and so forth, e-records management challenges, and policy and managerial implications of records management for SMEs. The chapter's intention is to showcase the importance of records management in general and e-records management in particular in business environments. Without effective records management, businesses cannot leverage their intellectual capital to effectively compete in the digital age. Poor records management, especially in governments, also erodes business confidence and discourages firms from investing in such jurisdictions because of delays experienced either in setting up businesses or litigation resulting from poor or a complete lack of documentation or documented procedures.

Chapter 11
Local Content and SMEs

INTRODUCTION

There are diverse opinions in literature on the meaning of the term 'content' going by the different definitions encountered in the course of writing this book. The lack of unanimity on what constitutes 'content' suggests that the definition of 'local content' is most certainly not cast in stone. This chapter first begins by unpacking the concept of content, before proceeding to local content, and how they can be applied to enhance the business values of SMEs in the global digital economy. Some literature tends to suggest that if an idea, or information, knowledge or data, is not in digital format, then it is not content. Siemens (2003) falls into this category, believing content to consist of e-journals, images, graphics, videos, movies, websites, online databases, emails, online news, software, and animations, among others. Claiborne (2005) simply defines content as the "stuff on your site". Some definitions do not differentiate between the media and content, while others perceive

DOI: 10.4018/978-1-60566-420-0.ch011

content to be recorded information or knowledge. Other definitions gleaned from the Web follow: content is everything included in a collection (WordReference, 2008); content refers to material that is of interest to users, such as text, images, music and movies (The Linux Information Project, 2005); content refers to information and experiences that may provide value to an end-user/audience (Wikipedia); and content is published information and experiences found in novels, movies, music, games, web pages, presentations, organised data, etc (Wiktionary). In their diversity, these varied definitions provide a framework for redefining the term 'content' on the one hand and 'local content' on the other.

From the definitions provided, one can notice a shift from the previously held notion that the format has to be digital in order for the object to be considered content. This notion may have perhaps arisen out of the tremendous impact of ICT and the Internet on content creation and distribution processes since the 90s, when the Internet revolution took off. During the last decade, there have been growing calls to generate content and make it available through new media in order to empower communities and create an inclusive knowledge society (UNESCO, 2005). Thus, in latter day definitions of content, there is less emphasis on the media or conduit (container) and more on the 'containee'. It is now more stringently acknowledged that the process of generating content (product, process or service) inevitably involves human or intellectual effort. What is 'content' must also be of use to the end-user.

It is important when redefining content to exercise caution so that media or conduits are not confused with the term. For example, a computer is not content, but the data stored in it is; a mobile phone is not content, but short message services (SMSs) are; and a book and website are not content, but the information contained within them are. Similar interpretations can apply to blogs, social networking sites, digital libraries, digital repositories, TV, radio, newspapers, etc; which are often perceived as content, when in fact they are media or conduits.

Consequently, 'content' is defined in this chapter as an intellectual process that results in a product (e.g. information, knowledge, website, song, dance, record, design, artefact, system, etc); process (e.g. engineering procedure, an algorithm, workflow, rules, etc); and service (e.g. public service broadcasting, news, advertisements, presentations, content scrapping, video on demand, animations, etc).

LOCAL CONTENT

The term 'local content' can be viewed as either a prefix of content or as its derivative. Consequently, like content, local content is a creation of the mind; in other words it is an intellectual product. UNESCO (2005) defines local content as the expression and communication of a community's locally owned and adapted knowledge and experiences within the scope of a specific environment. A community, according to this definition, is signified by location, culture, language, or area of interest. Such a community can comprise a whole region, a sub-region, a nation, a village, or a group of people with strong cultural, linguistic, and/or religious ties. Ballantyne (2002) perceives local content to be the proportion of goods and services produced locally, i.e. the development of local skills, technology transfer, use of local man-power, and local manufacturing.

The two definitions proffered above have implications for the scope of what constitutes 'local content'. Ballantyne's definition is broad-based and goes beyond information products (e.g. information, knowledge, data, etc), which often tend to be equated to content. This definition is, however, consistent with what Ugwanyi (2003) seems to portray when he observes that in Nigeria, the local content of goods and services in the oil and gas sector is less than 5%, and about 95% of the yearly expenditure (of about US$8 billion - N880 billion) flows out of the country through technical services and goods procured from outside the country. ComDev (2008), echoing Ballantyne's definition, notes that local content typically refers to company expenditure in the procurement of local goods and services, community development, direct hiring or employment, and building local skills. In contrast, the UNESCO definition intimates a close link between Indigenous Knowledge (IK) and local content. IK is an intricate knowledge acquired over generations by communities as they interact with the environment (Seepe, 2001). IK is also perceived to be a complex set of knowledge and technologies existing and developed around specific conditions of populations and communities indigenous to a particular geographic area (National Research Foundation, 2000). In this book, IK is taken to be a special form of local content characterised by the fact that it is based on experiences; often tested over centuries of use; adapted to suit local cultures and environs; dynamic and changing; oral-based, and acquired and passed down over generations; and is in the public domain and shared among community members. The first definition of local content as mentioned above (Ballantyne, 2002) does not seem to reflect much of a relationship with the concept of IK. However, the second definition (National Research Foundation, 2000) would seem to accommodate some aspects of IK, such as 'knowledge', 'technologies', and a particular geographic area (or community).

Based on these definitions, the following can be said to be examples of local content: local web content, IK, locally generated skills, locally generated services, local telephone directory entries, content of theses and dissertations, e-government content, grey literature content, national or community media content, locally generated songs and dances, organisations' records, workflows, national constitutions, personnel management systems, work procedures, community information, mortgage information, an amenities list, housing details, and much more.

Local Content Development Initiatives

Most international efforts aimed at developing local content focus on developing countries. This is perhaps because of the difficulties of accessing content in these countries, attributed largely to the digital divide. For example, Africa, which suffers a greater digital divide with the developed world than any other continent on earth, has witnessed an increase in national, regional and international efforts in local content development and documentation over the last two decades (Chivhanga, 2000). Several initiatives by development agencies, national governments and institutions are involved in these initiatives. For example, the Technical Centre for Agricultural and Rural Cooperation (CTA) has, since 1986, supplied major agricultural databases on CD-ROM to developing countries, including those in African institutions, to promote awareness of the benefits of IT and provide access to the world's agricultural research.

The G8 Digital Opportunities Task Force, in its 2001 Genoa Plan of Action, defined an approach to increase the access and use of ICT in countries worldwide. In particular, the Plan proposed developing local content by making software applications available, encouraging the participation of local stakeholders, and expanding the language base available on the Internet (D. O. T. Force, 2001). During the same summit, Canada undertook to make a contribution to the D. O. T. Force agenda for Africa by declaring its intent to promote community access through network connectivity and support for the creation of local content (Bourassa, 2003). Consequently, the Canadian based International Development Research Centre [IDRC] initiated telecentres in several African countries including Zimbabwe, Uganda and Swaziland (Graham, 1997).

The World Bank, in turn, has mainstreamed IK into its development processes in sub-Saharan Africa in order to optimise the benefits of these developments to the poor. The bank uses the Web to disseminate IK, and has a referral database. In Burkina Faso and Uganda, the World Bank has improved the connectivity and networking capacity of local communications. As a result, rural community telecentres are now used to obtain information from local communities and disseminate the knowledge generated to other regions. The World Bank is also involved in funding efforts to collect documents and disseminate IK (Naur, 2001). In 2002, the UK Department for International Development (DFID) started a project to support the development and exchange of local content in Africa. Specifically, the project aimed to support efforts by poor people in developing countries to create and exchange useful content (as perceived by these people) via ICTs (International Institute of Communication and Development, 2002). The project was simultaneously implemented in Kenya, Uganda, Tanzania, Mali and Zambia. A key aspect was the documentation of local content lessons for others to learn from. The Open Knowledge Network (OKN) arm of the project is the channel through which to disseminate and exchange the content generated in various media and in local languages. The project which is still running in the various countries, generates local knowledge products like CD-ROMs, video-tapes, leaflets, etc (International Institute of Communication and Development, 2002).

The World Summit on Information Society (WSIS), in 2003 and 2005 respectively, provided a platform for international debate on how to create an inclusive information society (Souter, 2007). Part of the WSIS put emphasis on the creation of local content. The WSIS Declaration of Principles and the Plan of Action noted that public institutions, such as libraries and archives, museums, cultural collections and other community-based access points, should be strengthened to promote the preservation of documentary records and the provision of free and equitable access to information (World Summit of Information Society, 2005). Likewise, the International Federation of Library Associations (IFLA), in its submission to WSIS, recognised the extent to which IK needs to be harnessed in the information society, and underscored the importance of plugging villages into the global network and establishing community access points; connecting universities, colleges, secondary schools and primary schools; creating scientific and research centres; and installing ICT in public libraries, cultural centres, museums, post offices, and archives. IFLA stressed the need for special attention to be paid to rural and under-serviced areas (International Federation of Library Associations, 2003).

UNESCO (2005), through its Local Content in the Media programme, has been involved in boosting the production and dissemination of local content using both traditional and new media in the most disadvantaged communities in the developing world by training content creators, supporting local content production, and enhancing local content distribution channels. In particular, UNESCO has helped by: training the creators of local content; producing local content; distributing local content; developing an audiovisual e-platform; and creating a content photobank. UNESCO's support is directed towards regional, interregional or international training initiatives, where the transfer of know-how between sub-regions or within a region is encouraged. Training areas focus on the chain of content production, including pre-production work, scriptwriting for television and radio, digital camera and sound techniques, digital editing, post-production, and self-training through production screenings. The training modalities require the development of guidelines, training modules and reference materials; publication of training materials; production of video tutorials; organization of short-term training courses; hands-on workshops; provision of experts for production screenings; and research on the latest trends and flows in audiovisual production.

The importance of local content was given a boost in 1980, in the Organisation of African Unity's Lagos Plan of Action. The declaration of the plan placed great weight on locally generated information's ability to solve Africa's development problems. It further recommended that African countries should enact policies that adequately reflect socio-cultural values in order to reinforce cultural identity. Moreover, the Plan stressed the importance of basic industries and the need to move towards greater technological capability and self-reliance in the region (Onyango, 1996).

African countries, through the e-Africa Commission (the ICT arm of New Partnership for Africa's Development - NEPAD), are making attempts to develop local content programmes. The e-Africa Commission aims in part to develop local content that would empower and propel Africans to uniquely contribute to the information society (Harbi, 2003; Ford, 2001; Mikenga, 2005). National projects aimed at local content development are therefore increasing in Africa. Cell phone service providers, for example, are beginning to exploit the potential of cellular handsets for local content publishing. Mobile phone content is becoming as pervasive as radio broadcasts to many African people living in rural areas. Southwood (2007) observes that for those living in rural and remote parts of Africa, cellular networks often provide the only way for people to make calls or send data because of the dearth of fixed-line phones. In 2007, it was estimated that almost 70% of the population in sub-Saharan Africa was covered by a cellular network. Local content that can be accessed through mobile phones includes news broadcasts, music, video, short message services (SMSs), video-conferencing, and Internet services, such as instant messaging and VOIP applications.

E-government projects are also emerging to play a leading role in promoting the creation and dissemination of local content. E-government is largely a local content intensive environment that consists of decision support systems such as records management systems, integrated financial management systems, human resource management systems, communication systems, and databases and portals (International Records Management Trust, 2004). E-government provides information services that are of immediate local use to its citizens and is therefore a rich source of local content.

CHALLENGES OF LOCAL CONTENT DEVELOPMENT

Various challenges face the development and management of local content, especially with regard to IK. In the developing world, IK has suffered from the lack of documentation, storage and retrieval mechanisms associated with conventional information systems. This is because IK is stored in peoples' heads and hearts, and is retrieved and communicated in temporal ways such as song, dance, plays, drama, story telling, etc. For decades, IK has remained clouded and obscured by western-schooled intellectuals, who often view such knowledge as backward and barbaric. Even today, despite its wider application, there is still limited awareness of the benefits of IK. The absence of laws that govern the practice and use of IK demonstrate, more than anything else, how governments in the developing world have contributed to the marginalisation of IK in favour of conventional knowledge.

Many challenges are associated with local content development and use, such as the limited financial resources available to developing countries for content production; inappropriate or insufficient training opportunities for content creators; lack of access to advanced technology and IT facilities, such as digital cameras and digital studios; low motivation and commitment at decision-making level to institute change; forces that do not encourage diversity; and inadequate intellectual property rights' provisions, because local content creators are not sufficiently recognised in IPR systems (UNESCO, 2003).

The lack of local content in the developing world is evident across all media and information channels. This is most keenly demonstrated by the overwhelming presence of content on TV screens coming from content providers in developed countries, reflecting languages, values and lifestyles that are often vastly different from those of the community consuming the content. The agencies that push global or non-local content are more powerful and resourceful than those disseminating local content (UNESCO, 2006). The challenges associated with local content development are primarily associated with sustainability, and in particular include (ComDev, 2008):

- Abrupt changes in external factors, such as in commodity prices, political transitions, or a project's characteristics. This can result in substantial social and economic disruption.
- Dependence arising from the failure to diversify the local and regional economy, leading to sole reliance on an individual company or sector.
- Disincentives, especially when contractual local content targets become unsustainable and are no longer flexible, carefully crafted, and/or regularly revisited.

- Cost escalations, especially if there is overinvestment in local content, which drains company resources away from core business.

The advent of globalisation, driven by international politics, advancements in technology, and the mobility of human capital across borders, both provides opportunities for the developing world, and threatens the foundation of IK. Part of this has been because globalisation has led to the deterioration of bio-cultural diversity as a result of western-led bio-prospecting activities. As a phenomenon, globalization also leans heavily in favour of developed nations due to overbearing western media content, sophisticated technological capacities, aggressive multinational conglomerates, superior scientific innovations, and restrictive intellectual property regimes. Consequently, the developing world's capacity to compete in the global knowledge environment is, in effect, diminished. Menchu (2007) acknowledges that although reliance on modern technologies has led to improvements in communication and agriculture, social cohesion, and knowledge and values; most of those they are intended to serve have been marginalised. Ballantyne (2002) observes that the powers that 'push' global or non-local content are often much stronger than those 'pushing' local content, as reflected by the over-reliance on foreign technical assistance, television programming, advertising, global brands, foreign languages in schools and universities, etc.

The application of ICT, especially the computer, has not been re-engineered for effective assimilation at grassroots level among communities that are not highly literate. People in economically and socially marginalised communities have been found to spend an inordinate amount of time and energy seeking and managing information related to survival and security (Agada, 999). Some work done in the US in 2001 on the subject of the digital government revealed how the lack of appropriate access points amongst communities hinders the provision of social services, for example by forcing individuals, often the poor, to travel long distances between offices (Bouguettaya et al., 2001). In Africa, the situation is even worse because of the lack of universal access to ICT facilities, especially among rural people. A study carried out in rural Kenya by Prof Souter (cited by the Kenya Times, 2005) on the *Economic Impact of Telecommunications on Rural Livelihoods and Poverty Reduction* found that although the Internet is increasingly available through public service points in many rural areas, it is scarcely used. The findings also showed that although the Internet is available in many local post offices and commercial outlets, such as cyber cafes, less than 2% of the people interviewed thought that the Internet is an important tool for passing on or receiving information.

Developing nations suffer from limited global Internet bandwidth. It is estimated that 98% of Africa's access to global Internet bandwidth goes through North

America, a condition that forces countries in Africa to incur high costs (with payment in US or Canadian dollars), leading to cash flows in directions wholly counter to development (Gerhan and Mutula, 2005). Fritz (2005), citing Missen, observes that Africa is a disconnected continent for many reasons. Telephone penetration is low, sometimes as low as 20 phones per 1,000 households. Furthermore in many instances, dial-in users share an Internet connection that is slower than their modem speed. On a tour of Africa in 2007, Craig Barret, Chairman of Intel, commented on the poor state of the infrastructure, noting: "We're just starting to see the Internet take off, but the lack of broadband capability is a real retarding force on the Internet in Africa" (Barret, 2007).

Prospects of Local Content for SMEs

Access to local content has important implications for SMEs wishing to effectively participate and compete in the digital environment. Without access to local content in the form of business information, prices, market opportunities, business trends, etc; it becomes difficult, if not impossible, to compete effectively against those with an abundance of such content.

Many communities in different countries have, at one point or another, relied on local content in the form of resources, values, and knowledge for their survival. Such local resources were developed and shared to strengthen the resource base of local populations, enhance their ability to integrate selected external elements into local practices, and broaden the options available to the people without diluting their local views and practices (Menchu, 2007). Local content provides opportunities for members of the community to interact and communicate with each other, and express their own ideas, knowledge and culture in their own language (UNESCO, 2005). Locally generated content is also more relevant to the consumer, and is easily accessible, cheaper, promotes community and cultural identity, enhances social inclusion, and facilitates universal access. Local content enhances economic development through the local procurement of goods and services, employment, and other related initiatives. It is therefore a significant factor in calculating overall competitive advantage, expanding global communication networks, enhancing transparency, and broadening stakeholders' awareness of social issues (ComDev, 2008).

Through local content, local communities can receive immediate benefits such as employment, and long term benefits such as skills transfer through training and direct experience. In addition, successful local content programs create a multiplicit effect that can help broaden and diversify the local economy. For example, the most important company benefit derived from a successful local content program is the social license to operate, as this helps to ensure uninterrupted operation and enhances the company's reputation. Other benefits of a local content program may include long term cost savings because of hiring locally instead of paying for staff relocation, and the reduced transportation and logistical expenses associated with local content arrangements (ComDev, 2008).

Local content draws on resources in the immediate vicinity and makes people less dependent on outside supplies, which may be costly, scarce and irregularly available. Moreover, local content satisfies internal needs, enhances self reliance, helps bridge intra and extra divides, enhances community access to content, and generally characterises a community because it mirrors real life situations and operations. IK (both tacit and explicit) as a specialised form of local content has been applied to fast-track medical experimentation through the use of local knowledge to find useful medicinal herbs and plants, and also in conflict resolution and environmental conservation through knowledge of indigenous communities. Local issues continue to matter to people, and digital local content could, in this respect, deliver a range of benefits, including more relevant local news; improved access to local services; greater capacity for individuals and local organisations to create and distribute their own content; support for local production and training; and improved advertisers' access to local markets (The UK Office of Communications Ofcom, 2006).

Policy and Managerial Responsibilities in Local Content Development

The increasing momentum of developing and disseminating local content requires a more systematic approach. The European Commission (European Commission, 2005) suggests that digital content programmes are needed to ensure that content is more accessible, usable and exploitable in a multilingual environment. Key steps for creating successful local content programmes at country level include (ComDev, 2008): encouraging a predictable regulatory and institutional environment that rewards investment in local content; integrating local content into broader project and community development planning; and conducting a baseline needs and capabilities assessment with key stakeholders in order to understand community priorities and identify existing skills. Other actions include:

- Establishing initial short and long term targets that are realistic, and developing partnerships with government or civil society to share accountability for meeting these targets and measuring success
- Creating decision making processes and partnerships that maximise success in order to resolve conflicting priorities, address performance issues, and meet local community expectations

- Monitoring and evaluating local content initiatives through stakeholder participation in order to improve chances of success, e.g. engaging with stakeholders to improve data collection.

Governments can stimulate local content growth by providing tax breaks on production equipment and other materials, and also encouraging co-production between foreigners and locals. Developing and nurturing local content requires multi-pronged approaches from national governments, with help from the international community. It is important for governments, industry, and researchers to work together to find how best to utilize existing infrastructures (e.g. media, ISPs, cell phone providers, etc) in order to successfully document and preserve the rich content of local knowledge resources. The private sector should be encouraged - through social responsibility - to fund community initiatives aimed at creating and documenting local content. The digitization of documents and creation of electronic databases that would be of use to local communities should also be encouraged. It is important to invest in the training of information professionals and develop skills in web design and the digitization of documents. It is also important to set up and maintain information gateways using software that addresses different languages, while also strengthening copyright and intellectual property laws.

The success of different initiatives in enhancing content development and documentation would depend on interventions that include: promoting literacy; providing diverse interfaces to document, store and access content; offering online education and support; promoting user-centred methodology in building systems; providing high bandwidth for multimedia content; building institutional capacities to collect, organise, store and share information and knowledge; stimulating the creation, processing, and dissemination of information to all people; providing local content online; and integrating ICT in the cultural milieu of the people (Woherem, 1993). Enhancing local content development would involve putting in place effective 'push' mechanisms; increasing and improving the supply of content; focusing on the demand side so that local content is more highly valued; and improving the packaging of content so that the content is more attractive and accessible.

CONCLUSION

Local content is important for SMEs, especially in terms of access to local human resources, technology, capital and know-how. SMEs largely rely on products and services in their immediate vicinity for their business transactions, and it is therefore important for requisite local content to be available on tap. Without access to local content, SMEs are forced to order or use products and services from elsewhere, and these may be expensive, unavailable and not suited for their work. Such SMEs will lack the capacity to offer a unique contribution to the international market that would enhance their competitiveness. In many instances, and relative to larger firms, SMEs find it more difficult to find e-business content that is applicable to them because of lack of time and relevant information and knowledge. Local content would increasingly empower SMEs' participation in the knowledge economy by facilitating connectivity; helping SMEs create and deliver products and services that are unique on a global scale; and providing access to new markets and new sources of competitive advantage that boost income growth. Policy makers must be aware of the opportunities and challenges of local content as a first step in promoting its development and use by the SME sector. Governments and the private sector have important roles to play in facilitating content availability across all platforms and encouraging the local development of new content, including content from public sources. Governments must consider and strike a balance between the need to protect intellectual property rights and their interests in fostering access to information and new ideas in order to promote innovation.

REFERENCES

Agada, J. (1999). Inner-city gate keepers: An exploratory survey of their information use environment. Journal of Information Science, 50(1), 74–85.

Ballantyne, P. (2002). Collecting and propagating local content development: Synthesis and conclusions. Retrieved on September 24, 2007, from http://portal.unesco.org/ci/en/files/5578/10342670280collecting_and_propagating_local_dev_Content.pdf/collecting%2Band%2Bpropagating%2Blocal%2Bdev%2BContent.pdf

Barret, C. (2007). Bridging Africa's digital divide. *BBC News*.

Bouguettaya, A., Ouzzani, M., Medjahead, B., & Cameron, J. (2001). Helping citizens of Indiana: Ontological approach to managing state and local government databases. *IEEE Computer*.

Bourassa, R. (2003, April 14-16). *Digital opportunities for Africa*. Paper presented to the Networking Africa's Future Conference, Pilanesburgh, South Africa.

Chivhanga, B. M. (2000). *Internet use in Africa*. Retrieved on April 12, 2002, from http://www.egroups.com/group/africa_web_Content_owner

Claiborne, S. (2005). *What is content part I?* Retrieved on October 26, 2008, from http://www.searchengineguide.com/scottie-claiborne/what-is-content-part-i.php

ComDev. (2008). *Local content in supply chain.* Retrieved on October 26, 2008, from http://www.commdev.org/section/_commdev_practice/local_content_in_supply_chain

European Commission. (2005). *E-Contentplus programme.* Retrieved on April 25, 2008, from http://ec.europa.eu/information_society/activities/eContentplus/index_en.htm

Force, D. O. T. (2001). *Digital opportunities for all: Meeting the challenge.* Retrieved on April 22, 2004, from http://www.dotforce.org/reports/DOT_Force_Report_V_5.0h.html

Ford, H. (2001). *African content in the multimedia age.* Retrieved on October 27, 2003, from http://www.eisa.org.za/category/technology/Content1.htm

Fritz, M. (2005). *Food for thought: Localized content aids disconnected Africa.* Retrieved on April 24, 2008, from http://www.eContentmag.com/Articles/ArticleReader.aspx?ArticleID=8261

Gerhan, D., and S. M. Mutula (2005). Bandwidth bottlenecks at the University of Botswana: Complications for library, campus, and national development. Library Hi Tech, 23(1), 102–117. doi:10.1108/07378830510586748

Graham, M. (1997). *Use of information and communication technologies in IDRC projects: Lessons learned, IDRC study/Acacia initiative.* Retrieved on November 13, 2002, from http://www.idrc.ca/acacia/outputs/op-eval3.htm

Harbi, M. (2003). *New partnership for Africa's development (NEPAD): A bold initiative for Africa's revival.* Retrieved October 26, 2003, from http://ww.itu.int/itunews/issue/2003/02/partnership.html

International Federation of Library Associations. (2003). *Libraries the information society in action.* Retrieved on February 14, 2006, from http://www.ifla.org/III/wsis/Byrne-Plenary-Address.html

International Institute of Communication and Development. (2002). *IICD supported project: Open knowledge network (OKN).* London: DFID.

International Record Management Trust. (2004). *The e-records readiness tool.* London: IRMT.

Kenya Times. (2005). *Economic impact of telecommunications on rural livelihoods and poverty reduction: A study of rural communities*. Nairobi: Kenya Times.

Menchu, R. (2007). *Learning endogenous development: Building on biocultural diversity*. Bourton on Dunsmore, UK: Intermediate Technology Publications Ltd.

Mikenga, S. (2005). *Africa's first NEPAD e-school is launched in Uganda*. Pretoria: Nepad e-Africa Commission.

National Research Foundation. (2000). *Notes on the nine research focus areas*. Retrieved on November 20, 2008, from http://www.nrf.ac.za/publications/news@nrf/july2000/fanotes.stm

Naur, M. (2001). Indigenous knowledge and HIV/AIDS: Ghana and Zambia. IK Notes, 30, 1–4.

Office of Communications (Ofcom). (2006). *Ofcom examines prospects for local digital content services*. Retrieved on September 25, 2007, from http://www.ofcom.org.uk/media/news/2006/01/nr_20060119

Onyango, R. (1996). Indigenous technology capacity: Can social intelligence help: A Kenyan case study. In B. Cronin (Ed.), *Information development and social intelligence* (p. 164). London: Taylor Graham Publishing.

Seepe, S. (2001). Indigenous knowledge systems. *Mail & Guardian*, 22.

Siemens, G. (2003). Content management*: Our organized future*. Retrieved on October 17, 2006, from http://www.elearnspace.org/Articles/Content management.htm

Souter, D. (2007, September 17-19). *Whose summit? Whose information society? Civil society and WSIS*. Paper presented at the Centre for the Study of Globalisation and Regionalisation 10th Anniversary Workshop on Digital Divide, A Legal Postscript to the World Summit on Information Society, University of Warwick, Warwick.

Southwood, R. (2007). Kenya: Country makes a technology leap with mobile TV. *Balancing Act News, 16*.

The Linux Information Project. (2005). *Content definition*. Retrieved on October 26, 2008, from http://www.bellevuelinux.org/content.html

The Standard. (2004). *Enact policy to protect indigenous knowledge. Nairobi: The Standard Ltd. Translate.org.za.2003. Open source software translation project*. Retrieved on May 5, 2003, from http:/www.translate.org.za Ugwuanyi, E. (2003). FG loses N880b annually to capital flight in upstream activities. *Vanguard*. Retrieved on October 27, 2003, http://fr.allafrica.com/stories/200310080771.html

UNDP. (2007). *The role of governments in promoting ICT access and use by SMEs*. Retrieved on November 18, 2008, from http://www.apdip.net/apdipenote/12.pdf/

UNESCO. (2003). *Access to knowledge, open access, cultural and linguistic diversity*. Local paper presented at WSIS PrepCom II Multistakeholder Roundtable No. 2. Retrieved on October 26, 2008, from http://www.itu.int/wsis/docs/pc2/roundtables/rt2/khan.pdf

UNESCO. (2005). *What is local content?* Retrieved on October 26, 2008, from http://portal.unesco.org/ci/en/ev.php-URL_ID=4237&URL_DO=DO_TOPIC&URL_SECTION=201.html

UNESCO. (2006). *The need for local content*. Retrieved on October 27, 2008, from http://portal.unesco.org/ci/en/ev.php-URL_ID=5463&URL_DO=DO_TOPIC&URL_SECTION=201.html

Woherem, E. (1993). *Information technology in Africa: Challenges and opportunities*. Nairobi: ACTS.

WordReference. (2008). *Content*. Retrieved on October 26, 2008, from http://www.wordreference.com/definition/content

World Summit on the Information Society. (2005). *Tunis commitment and Tunis agenda for the information society*. Retrieved on February 14, 2006, from http://www.itu.int/wsis/documents/doc_multi.asp? lang=en&id=2266|2267

Chapter 12
Knowledge Management (KM)

INTRODUCTION

There are a growing number of studies on knowledge management (KM) in SMEs for various reasons. One has been the growing realisation that SMEs are in a unique situation because their most significant assets are intangibles comprised mainly of knowledge. SMEs constitute the largest number of enterprises in the economies of both developed and developing countries, as already discussed in preceding chapters. In Germany, for example, 97.9% of all companies fall within the scope of SMEs and provide approximately 36% of all industrial investments (Wimmer and Wolter, 2000). Likewise in Australia, SMEs account for 97% of all private sector businesses and produce 30% of the nation's output (Australian Bureau of Statistics-ABS, 2001). Handzic (2006) stresses that organizations that manage knowledge

DOI: 10.4018/978-1-60566-420-0.ch012

better will deal more successfully and effectively with the challenges of the new business environment. KM is therefore perceived to be a key factor in realizing and sustaining organizational success for improved efficiency and innovation.

Although there has been a lot of literature on the practise of KM within large organisations, the same cannot be said for SMEs. Rasheed (n.d) argues that this is because in SMEs, the managers are also in most cases the owners, which implies that decision-making is centralized and that fewer layers of management exist. Decision-making processes are therefore shorter in SMEs than they are in large organisations, and therefore KM practices are not prevalent. However, the owners of SMEs have the opportunity to become the key drivers of knowledge management implementations in their organisations. SMEs have an advantage over larger enterprises with regard to KM because they have a simpler, flatter and less complex structure. This kind of structure can comfortably facilitate change across the organisation because functional integration, both horizontally and vertically, is easier to achieve, and fewer complications are likely to arise. Comparatively, larger organisations have a bureaucratic structure, making them slower and less flexible in adopting new schemes. However, larger organisations also have the advantage of specialization in their staff roles, which gives them better expertise in implementing knowledge management. In terms of the culture of organisation, SMEs tend to have a smaller number of people, usually united under common beliefs and values, which implies that it is easier for them to change and implement knowledge management. Generally, it is therefore easier to create a knowledge sharing culture in small-sized enterprises than it is in larger ones.

Knowledge Management Benefits and Applications to SMEs

Blogie (2007) underlines the benefits of KM for SMEs; top among them is the fact that organisational knowledge has become the most significant source of competitive advantage for businesses as a result of the rapid expansion of goods and factor markets. SMEs compete with limited physical assets and are thus forced to maximise their utilization of knowledge, which is their most abundant internal resource. In the global arena, where better access to external resources is expected, organisational knowledge is identified as a strategic asset that is not easily imitated by the competitors, thereby providing a sustainable edge for the firms that are aware of and use it (Bollinger and Smith, 2001). This means that in order for SMEs to succeed in their respective markets, they must ensure that they have control over their internal knowledge.

Knowledge management is an important tool in any organization, especially in business enterprises, and inevitably managers are expected to understand, appreciate and adopt KM practices. Within the evolving digital economy, forces are emerging that have the potential to reshape the economic and organisational environment. These forces include globalisation; the emphasis on product and service quality; new technology; changing economic and political structures; deregulation; the competition for the dollar and euro; and a new breed of sophisticated customers who have changing requirements and expectations (Rasheed, n.d). These driving forces have led individuals and organisations to appreciate the important role of knowledge in an increasingly competitive global market. KM is not a practice established for the sake of academics only; rather it is a realization that knowing about knowledge is critical to business growth and survival.

The benefits that accrue from applying KM practices may vary from organisation to organisation and are based on the orientation of their operations. Nevertheless, the primary business uses or domains of KM include (Dyer and McDonough, 2001): capturing and sharing best practices, providing general and corporate training, managing customer relationships, delivering competitive intelligence, providing project workspaces, managing legal and intellectual property, enhancing web publishing, enhancing supply chain management, etc. Generally, companies that adopt KM enjoy the following benefits: improved profits, growth in revenues, support for e-business initiatives, and shortened product development cycles. Organisations that reward collaboration and information sharing are outperforming companies that discourage these practices as a result of enhanced productivity, increased competitiveness and lowered costs (Microsoft, 2001). KM enables knowledge intensive organizations to effectively manage their knowledge resources; harness the knowledge of aging employees and departing staff in order to preserve institutional memory; and facilitate capacity building plans in knowledge management and adaptation to improve technology transfer within organizations.

With regard to the administrative services in organisations, KM can be used to develop portals for financial services (e.g. budgeting and accounting, best practices, procedures, templates, procurement - purchasing, warehousing, etc) and human resources management (e.g. vacancies, payrolls, affirmative action, best practices, templates, etc). In terms of strategic planning, KM can be used to compile a database of internal information that catalogues strategic plans and provides reports from external environmental scans, competitor data, and links to research groups. Benefits here would include an improved ability to support the trend towards decentralised strategic planning and decision-making; and the improved sharing of internal and external information (Kidwell et al., 2000).

Within disparate businesses, KM can be used to help connect the 'silos' that hamper communication with the other units of the organization. The successful application of KM practices raises the awareness of leaders and managers, and also of frontline personnel, of the advantages that KM can bring to an organization. At the individual level, KM practices provide opportunities to employees for career enhancement and development.

Knowledge Management Concept

Knowledge management, the process of creating, storing, sharing and re-using knowledge, enables organisations in both the public and private sector to achieve their goals and objectives. The essence of knowledge management is to provide strategies that get the right knowledge to the right people at the right time and in the right format. Knowledge is the basis of the knowledge management process. Such knowledge can be explicit (documented information expressed in formal shared language) or tacit (embodied in peoples minds). Three levels of knowledge management exist, namely the track, organisation, and individual levels. The track level of knowledge management can be sub-divided into two categories, i.e. IT-Track knowledge management (management of information) and People-Track knowledge management (management of people). At the IT-Track KM level, researchers and practitioners are mostly trained or educated in computer and/or information science. They are involved in the construction of information management systems, artificial intelligence, reengineering, groupware, and the like. To them, knowledge is equated to objects that can be identified and handled in information systems (Sveiby, 2001). Although the IT-Track Management level is still novel, it is growing rapidly because of the influence of IT.

At the People-Track level of KM, researchers and practitioners mostly have backgrounds in philosophy, psychology, sociology, business or management. They are primarily involved in assessing, changing and improving individual human skills and behaviour. The People Track level is also more about how to maximise the capabilities of human resources by encouraging individuals to create new knowledge, share it and be more innovative. To them, knowledge is equated to processes, know-how, or a complex set of dynamic skills that are constantly changing. Both researchers and practitioners are traditionally involved in learning and in managing these competencies individually. Unlike the IT-Track level, this track has matured and its growth is almost levelling off.

The Individual level's focus in research and practice is on the individual, while the organisational level's focus is on the organisation. The organisation's role here is to provide an environment that would nurture the sharing of knowledge and allow the staff to try new things, as this may result in new products, new markets or the acquisition of competitive advantage.

Knowledge management is said to have evolved in corporate environments during the 1980s as a result of the revolution in ICTs. During that period, knowledge started to emerge as a strategic resource. Porter (1985) is believed to have originated the concept of KM when he coined the term 'competitive advantage' to describe a way of creating added value for customers by offering them unique products and services at cheaper costs. The KM concept was later taken up by Drucker (1987) when he wrote about knowledge workers, thus exposing the issue of managing knowledge (Organisation of Economic Cooperation and Development, 2003). Now, many organisations, especially in the corporate sector, are investing heavily in the development of knowledge management systems to support knowledge work and enhance organisational learning (Davenport and Prusak, 1998). Due to earlier initiatives, the private sector can boast of having the most advanced knowledge management systems.

Knowledge management focuses on the following aspects (White, 2004):

- Intellectual capital or intelligence
- Personal experiences
- People's beliefs, perspectives, judgments, expectations, intentions, relationships, etc
- Know-how, skills and competencies
- Learning, critical thinking, and innovation
- Knowledge application
- Explicit and tacit representations.

Knowledge management also involves critical thinking; innovation; the ingredients of originality, adaptability, intelligence and learning; and competencies and the sharing of experiences. KM can be characterised as being reliant on modern technologies that cut across cultural and traditional structures (Grey, 1998).

KNOWLEDGE MANAGEMENT STRATEGIES

Hansen et al. (1999), in a study on the knowledge management strategies of companies in several industries, particularly consulting firms, encountered two major practices of knowledge management that they respectively referred to as codification and personalisation. In the codification approach, information is carefully codified and stored in databases where it can be accessed and easily used by anyone in the company. Codification involves using a 'people-to-document' approach where knowledge is extracted from the person who developed or found it, made independent of that person, and thereafter reused for various purposes. After extracting knowledge from its compiler, knowledge objects are developed by extracting key components of knowledge, such as interview guides, work schedules, benchmark data, and market segmentation analyses; and storing them in the electronic repository for people to use. This allows many people to search through and retrieve codified knowledge without having to contact the person who originally compiled it. This approach makes it easy to reuse knowledge, particularly when developing project proposals, and thus saves a great amount of time.

Personalisation in knowledge management refers to knowledge being closely tied to the person who developed it, and is mainly shared through person-to-person interaction (Hansen et al., 1999). This strategy focuses on the dialogue between individuals and not on knowledge objects in a database. In this case, knowledge that has not been codified is transferred in brainstorming sessions and in one-to-one conversations. To make this knowledge management strategy work, heavy investment is made in building networks that connect experts. Knowledge is shared not only in face-to-face meetings, but also over the telephone, through e-mail, via videoconferences, etc. Networks are developed by moving people between offices, promoting a culture in which consultants are expected to promptly return phone calls from colleagues, creating directories listing experts, and using consulting directors within the firm to assist project teams.

Firms have been found to focus on one of the strategies and use the other in supporting roles. Hansen et al. (1999) state that implementing codification creates high quality, reliable and fast information systems because of the continuous reuse of information. The information systems invest once in a knowledge asset and reuse it several times thereafter. However, codification invests heavily in IT systems, while personalisation invests only moderately in IT resources. Codification is useful when an organisation is dealing with similar problems all the time. Personalisation is more relevant in a rapidly changing environment. Most solutions in codification have been proven and are reliable. The codification approach has been found to be cost-effective, especially in reusing software programmes. Personalisation, on the other hand, can be very expensive because of high consultation fees. Comparatively, codification also enhances the sharing of knowledge while the personalisation strategy relies on tacit knowledge, which is difficult to share and is time consuming, expensive and slow to transfer. If a company offers a standardised or mature product, it may benefit from the reusable model. On the other hand, if a company relies on tacit knowledge for problem solving, then the personalisation model may be more appropriate. Kidwell et al. (2000) provide the example of Nokia as a company that uses codification practices to make sense of market trends and customer requirements in order to quickly put that knowledge into action in the product development pipeline. Nokia Company's management system provides relevant information rapidly to whoever needs it in the organisation. It has a strong customer focus and service ethic, and is innovative in a highly competitive market.

Knowledge Management Practices: General Adoption Motivations

The growing impetus to adopt KM, especially in corporate organisations the world over, can be attributed to both external and internal factors. Agarawal (n.d) opines that pressures of the marketplace and increased competition are pushing [all organisations] to think like businesses. Moreover, markets are going global; thus organisations are attempting to internationalise and offer high-quality services to clients regardless of location. Organisations therefore have to adjust and develop strategies to respond rapidly to changes in technologies and the increasing demands of stakeholders. Furthermore, as the foundation of industrialised economies has shifted from natural resources to intellectual assets, executives have been compelled to re-examine the knowledge underlying their businesses and how that knowledge is used. Increasingly, the knowledge economy expects businesses to be part of knowledge generation, innovation and production processes. The dynamism within the trajectories of knowledge and technology is such that the skills to cater for specific needs are changing so fast that they increasingly have to be generated on the job. This means that knowledge origination, technology advancement and economic activity require persons with very high levels of generic knowledge.

Participation in the knowledge economy requires the ability to constantly renew economic and social systems and extend knowledge and specialist skills to engage effectively in knowledge production (Kok, 2007). To re-iterate, the digital age has made it possible to codify, store, and share certain kinds of knowledge more easily and cheaply than ever before. The increased external and internal pressure necessitates KM incorporation into organisations so that they (organisations) may optimise the use of the information systems in place; provide standard techniques for data collection; mitigate high staff turnovers; integrate technology into their operations, missions and visions; and build trust among staff with regard to the sharing of data.

The increasing adoption of KM, especially by the executives in corporate organisations, has also been inspired by the need to tap into world class business practices that are espoused by multinational companies. Another important factor relates to brain drain, which results in the massive net outflow of intellectual capital and the consequent need to explore technologies' application or other methodologies to facilitate the retention of these knowledge assets within organisations and beyond a person's physical departure (McLean, 1999). The transformation of economies, especially of those that are highly regulated, also requires rapid knowledge transference to improve the success of the transformation process into free market economies. In a knowledge-intensive economy, enterprises increasingly compete with each other for the use and provision of knowledge-intensive resources such as human capital and new technologies (Organisation of Economic Cooperation and Development, 2003).

Although the corporate sector is increasingly adopting KM practices, the same cannot be said of governments, which, until recently, have not been at the forefront of integrating knowledge management in their operational milieu. However, increasing e-government initiatives the world over have slowly made the public sector realise the strategic importance of knowledge (Wimmer, 2004). In Europe, for example, KM ranks high on the management agenda of most central government organisations. Knowledge management is also seen as one of the top five future internal management priorities in a number of organizations in OECD member states (Organisation for Economic Cooperation and Development, 2003). KM-driven economies in OECD countries recognise the need for efficiency and productivity; improving transparency and the outward sharing of information; and improving working relationships and trust within organizations. The KM approach in managing public sector knowledge resources promotes life-long learning, the sharing of knowledge across different units in organisations, and reaching out to the public. Knowledge management is also important in public sector transformation (such as e-government) because such transformation involves the emergence of a knowledge-enhanced government. Essentially, the knowledge economy is undergoing a structural transformation where the rapid creation of new knowledge and the improvement of access to knowledge bases are factors that increase efficiency, innovation, the quality of goods and services, and equity. The motivation for adopting KM in organizational operations was given impetus by the final outcomes of the World Summits on Information Society in 2003 (in Geneva, Switzerland) and 2005 (Tunis, Tunisia) respectively (International Telecommunication Union, 2005).

Knowledge Management Processes and Activities

Knowledge management entails the discovery, acquisition or creation of knowledge (through research); the transmission or dissemination of knowledge (through teaching); the application of knowledge to human problems in the interests of public service; and the preservation of knowledge in libraries, museums and archives (Allen, 1988:66). Bhatt (2001) identifies five phases of KM, namely knowledge creation, validation, presentation, distribution, and application. Knowledge creation refers to the ability of an organisation to develop novel and useful ideas and solutions. This phase represents an emergent process comprising of motivation, inspiration and experimentation. The knowledge validation phase is concerned with the extent to which a firm can reflect on knowledge and evaluate its effectiveness within the existing organisational environment. Knowledge presentation refers to the way knowledge is displayed to the organisation's members, i.e. whether in print, on CD, or through optical media. Phase four, knowledge distribution, represents the distribution and sharing of knowledge across the organisation and requires interactions between organisational technologies, techniques and people. Finally, knowledge application is concerned with the art of locating the right knowledge in the right format in any organisation.

KM consists of three major components, namely the people who create, share and use knowledge; processes that acquire, create, capture, organise, share, transfer and apply knowledge; and technology that stores and provides access to knowledge. Processes in KM may include scanning, document imaging, forms processing, classification, indexing, categorisation or taxonomy, backup and recovery, search and retrieval, publication, archiving, storage, migration, records management, e-mail management, and more (Wagner, 2008). Within an organisation, knowledge is embodied in the wisdom of people, expertise and talents, databases, records, portals, formal and informal meetings, memos, infrastructure, visions/missions, strategic plans, budgets, and human resource systems, to name a few. Technology is at the centre of knowledge management, and helps retrieve the various forms of information and knowledge embodied in systems, institutional repositories, processes, strategies, methodologies, e-mails, patents, products, services, etc; within and outside organisations. Organisations that are KM-intensive have some or more of the following knowledge-based resources or infrastructures (Corall, 1999):

- Knowledge databases and repositories (explicit knowledge)
- Knowledge route-maps and directories (tacit and explicit knowledge)
- Knowledge networks and discussions (tacit knowledge).

Knowledge databases and repositories consist of stored information and documents that can be shared and re-used, such as client presentations, competitor intelligence reports, customer data, marketing material, meetings' minutes, policy documents, price lists, product specifications, project proposals, research reports, training packs, and others. Knowledge route-maps and directories, on the other hand, are tools that point to people, document collections and datasets that can be used for consultation, such as the yellow pages, and 'expert locators' containing CVs, competency profiles, and research interests. Knowledge networks and discussions refer to networks that provide opportunities for face-to-face and electronic interaction, such as chat facilities to foster learning groups or hold best practice sessions.

Increasingly, the concept 'digital dashboard' has emerged as a KM tool customised for knowledge workers. It consolidates personal, team, corporate and external information and provides 'single click' access to analytical collaborative resources. It brings an integrated view of a company's knowledge resources onto an individual's desktop, thereby enabling better decision-making by providing immediate access to key business information (Microsoft, 2000).

With regard to KM processes, Skyrme and Associates (2002) outline the following:

- Knowledge discovery – data mining and text mining
- Content creation – creativity tools and content management systems
- Infrastructure – intranet, extranet and portal building software
- Information retrieval – search engines and intelligent agents
- Expertise profiling - expertise directories
- Knowledge sharing - computer conferencing, portals, and document management
- Integration –KM suites, enterprise application integration, portals and standards.

KM also provides tools, procedures and an environment for KM policy formulation and performance management; knowledge discovery, creation, capture and sharing; knowledge mapping and information architecture; the adoption and use of standards – metadata and best practices; developing communication systems; methodologies for content management and content quality management; developing information security plans; and facilities for integration, interoperability and scalability.

Knowledge taxonomy, also known as knowledge organisation, refers to the classification of knowledge assets in order to make them easier to find and understand (Herd, 2001). Knowledge taxonomies are being applied in customised search engines and as an integral part of web portals to enhance resource discovery. They are also important in facilitating the sharing of a common language for classifying knowledge; searching for and retrieving knowledge, especially with respect to controlled vocabularies in search engines, web content managers, and online databases; improving internal communication by developing common standards that can be understood by all; enhancing intelligence gathering; organising web portals to enhance the management and retrieval of information; providing networked environments for effective access to information; and making knowledge manageable, accessible and retrievable.

Knowledge packaging is an important activity in KM, and refers to the process of putting knowledge in a format that is necessary for its transmission and application. It can also refer to the methods, tools and techniques used to formalise experiences and know-how and transform them into products and services (Ardimato et al., 2006). Essentially, knowledge packaging entails creating products or services, converting them into useable formats, and subsequently commercialising them. This packaging process begins with intelligence, which is largely uncodified (implicit) knowledge, followed by codification (making it explicit), which may include developing processes, creating documents, making drawings, or creating databases, websites, publications, etc. Once the process of codification is over, the resultant products or services are diffused. Various formats or products exist for the packaging of knowledge, including (Srinivas, 2007) abstracts, alerts, announcements, policy briefs, bibliographies, indexes, catalogues, best practices, brochures, books, bulletins, charts, databases, diaries, blogs, commentaries, annotations, journals, metadata (i.e. hyperlinks), models, pamphlets, posters, slides, standards and practices, directories of expertise, intranets and portals, knowledge centres, domain know-how, extranets, expert systems and customer profiles, among others. The purpose of packaging knowledge is to facilitate its communication and enhance its understanding, commercialisation, and use.

Knowledge audit is another important aspect of KM. According to Hylton (2003), it is a comprehensive investigation into the entire knowledge-base of the organisation, resulting in a clear indication of how well the organisation's knowledge is being exploited in business activities and in alignment with the goals and mission of the organisation. It answers questions such as what knowledge exists and where it can be found; where and how it is created and who owns such knowledge; and what knowledge assets the organisation has and where they are. A knowledge audit should be undertaken before an organisation embarks on knowledge management activities. The knowledge audit process is people-centred because all employees within an organisation carry useful or important knowledge, skills and experiences about the organisation they work for. A knowledge audit identifies knowledge gaps and what knowledge is needed to support the organization's overall goals and individual and team activities; provides a map showcasing the flow of communication; and reveals good practices, barriers and blockages in the organization. A knowledge audit enables the organization to determine a strategy for developing KM initiatives and projects.

A knowledge map is a virtual tool that represents an organisation's knowledge. It is actually a navigational aid to codified information and tacit knowledge, showing the relationships between knowledge stores. Knowledge mapping encourages re-use and prevents the re-invention of information. It also helps when identifying knowledge sources, expertise and ways of building bridges to increase knowledge sharing. It reduces the burden on experts and supervisors by helping staff find information on their own. In order to map where real knowledge resides and its value, it is necessary to understand how it is built, maintained and used. Knowledge maps make corporate assets visible to all members of the organisation.

Knowledge Management Technologies

KM relies on ICT to facilitate knowledge creation, capture, organization and transmission. ICT ensures the accurate and timely exchange and delivery of knowledge in a way that is far more efficient than with the use of human actors. It is, in fact, impossible these days to disseminate and utilize information without incorporating ICT. ICT can facilitate KM in various ways (Nonaka and Takeuchi, 1995), for example:

- ICT represents the most convenient and cheapest way for individuals to share tacit knowledge with their colleagues
- ICTs such as the e-mail provide a platform for the quick and efficient exchange of information
- New web-based software and servers facilitate explicit knowledge sharing and also use intranet home pages for publishing applications
- Computer applications assist data mining using tools based on neutral networks, simulation modelling, and other applications that use visualization technologies, such as geographic information systems
- Using ICTs, many organizations have invested in the creation of large stores of data, such as data-marts or warehouses, in order to support the information needs of decision makers
- ICT is used to store, retrieve, disseminate and manage knowledge through databases and websites.

The various technologies that organisations use to leverage KM are largely dependent on the Internet and the World Wide Web. The rapid development in information technology and the convergence of telecommunications, computing and broadcasting, have been welcomed because of how they improve communications within and outside the organisation. This has been further enhanced by the great potential of the Internet and its associated new technologies in their ability to support a large number of users and devices and in new developments in bandwidth, content and applications. New technologies can broadly be grouped under search and retrieval, networking, hardware, and software. Technologies that have recently been growing in popularity include Wi-Fi, Wi-Max, Bluetooth and Voice Over IP (VOIP). These Web-based technologies are popular in knowledge management implementation because they are convenient to use, easy to develop and maintain, and provide one of the quickest and most far-reaching ways through which to convey information. Working closely with new technologies is Web 2.0, defined by O'Really (2005) as "the network platform, spanning all connected devices and applications that make the most of the intrinsic advantages of that platform, delivering software as a continually-updated service that gets better the more people use it; consuming and remixing data from multiple sources, including individual users, while providing their own data and services in a form that allows remixing by others".

Examples of Web 2.0 applications include blogs, wikis, content aggregators, and much more. Blogs provide authors with ownership of the medium of communication and allow instant publication (i.e. anytime-anywhere) of text or graphics on the Web without the need for sophisticated technical knowledge. They also provide the means by which people give comments or feedback to each posting on the blog. Blogs also allow users to archive past blog posts by date and provide hot links to other bloggers. Blogs therefore promote collaborative learning and the sharing of information. They serve as forums for announcements and as sources of news (often local). Wikis enable real time conversations and permit contributors to build knowledge management applications. They provide an environment for unique content creation. This requires content aggregators - individuals or organizations that gather Web content and applications from different online sources for reuse or resale - and programs or feed readers, which check a list of feeds on behalf of the user and display any updated articles that they find. Content aggregators enable users to subscribe to a feed by supplying to their reader a link to the feed. The user can then check the subscribed feeds to see if any of those feeds offer new content since the last time they were checked and if so, retrieve that content and present it to the user.

CHALLENGES OF KNOWLEDGE MANAGEMENT

Knowledge management presents a number of challenges to organisations. Maponya (2004) found that [organisations] do not generally manage information well - they are more likely to lose, duplicate, or fail to share or exploit information; do not always 'know what they know'; and do not recognise knowledge as an asset. Milam (2001) identifies the following knowledge management challenges: insufficient time for

KM (especially in terms of staff); the failure of the current culture of organisations to encourage sharing; poor understanding or awareness of KM's benefits; lack of skills in KM techniques; lack of funding for KM; lack of incentives or rewards to share knowledge; lack of appropriate technology; and lack of commitment from senior management. The field of KM is also still relatively young, and executives have therefore lacked successful models that they could use as guides. Most knowledge management systems' problems relate to inadequate or the outright lack of human resources, infrastructure, finance, literacy, relevant content, and technologies.

The challenges of KM also stem from problems encountered when handling knowledge in different formats; issues of protection - intellectual property and confidentiality; patent processing; dealing with intangible products; and auditing intellectual capital, which entails establishing its existence, its (intellectual capital's) ownership and its value. Moreover, knowledge management faces difficulties with (Storey and Quintas, 2001):

- Winning the trust, motivation and commitment of stakeholders
- Managing workers who are not conventional employees, such as contract staff and consultants
- Relying on knowledge workers who may leave the organisation without their knowledge having been captured
- Vulnerability of web-based systems and other technological infrastructure
- Replication and piracy.

POLICY AND MANAGERIAL RESPONSIBILITIES IN KNOWLEDGE MANAGEMENT

In designing KM solutions, companies need to think carefully about why they are needed. Findings from SME case studies indicate that it is important to translate the knowledge held by key personnel into organizational knowledge by embedding routine processes and procedural knowledge into standard operating procedures; and to codify implicit knowledge (Microsoft 2000).

The UN's 2008 E-government Report (United Nations, 2008) provides a good summary of what organisations need to do in order to implement KM practices. With respect to human resources, the survey stresses the importance of building an environment that instils trust among employees. This in turn implies the selection and development of leaders who promote information sharing. Effective knowledge sharing can be encouraged by rewarding those who input information into the system by establishing a formal structure of incentives and rewards. For management, information sharing is important because it results in reduced information costs. Employees also have embedded knowledge of the organization's values and objectives, so they must be encouraged to use their own creativity and innovation to turn their ideas into valuable products and services.

Leaders must put in place strategies that encourage creativity and innovation among their employees. They must also improve the use of knowledge that exists within the organization by sharing best practices. Organisations should also strive to enhance Customer Relationship Management (CRM), which revolves around enhancing customer focus and building relationships with private sector partners. This would be helped along if organisations' KM strategies address issues of confidentiality, data integrity and the availability of information. While confidentiality deals with the unintentional disclosure of information outside authorized parameters and data integrity assures the trustworthiness of the information, availability ensures that the information is made available at the request of authenticated clients.

For the effective management of knowledge resources within organisations, deliberate actions are needed by knowledge managers. Staff development, by sponsoring learning and training, is critical because it keeps staff well informed. Strategic plans should also be put in place to deal with recruitment and succession. The conversations between people inside the organization with customers and suppliers are crucial to sensing trends and emerging needs (Leonard, 1999). Staff should be encouraged to keep abreast of developments in the area of KM through research, conference attendance, or through consultations. Attempts should also be made, where applicable, to provide rewards and incentives to staff in order to recognise and give credit to those taking advantage of KM to improve their performance at work. Organisations also need to engage in continuous process reengineering and redesign in order to attract new knowledge from both internal and external sources.

Knowledge management can be enhanced in organisations by motivating staff so as to avoid brain drain and the depletion of the talent pool. To achieve this, organisations may offer the security of long term employment, which would also minimise high staff turnover and the costs of knowledge maintenance. There is also a need to maintain continuity and a recognizable identity over time, and be able to integrate and manage organisations' knowledge, information and their exchanges, both within and outside the organisation, for effective performance (Leonard, 1999). Therefore, organisations need to develop mechanisms that capture relevant knowledge and disseminate it accurately, consistently, and in a timely manner (Bollinger and Smith, 2001). Similarly, it is important for organizations to establish a culture that encourages the creation of new knowledge while improving the application, transformation and integration of existing knowledge within the organisation's KM systems. Kidwell et al. (2000) suggest the following steps for organisations contemplating the adoption of KM approaches:

- Start with strategy and determine what you wish to accomplish with KM
- Avail human resources, financial resources and information technology to support KM
- Seek a high-level champion who can argue for why KM is necessary
- Select a pilot KM project with high impact and low risk on the organisation
- Develop a detailed action plan for the pilot that defines the process, the IT infrastructure, and the roles and incentives of the project team
- After the pilot, assess the results and refine the action plan.

KM must be integrated into the firm's existing strategic direction so that core competencies and employee capabilities can be positively affected and the performance of the organisation improved. A knowledge audit is necessary before implementing KM approaches. Such an audit would facilitate findings, analyses, interpretations and reports on a company's information and knowledge policies, and its knowledge structure and knowledge flow (Hylton, 2002).

Oosterlinck's (1999) proposals on how [organisations] can improve KM practices include: drafting a mission statement that incorporates KM; creating awareness concerning the responsibility and accountability of the staff towards the stakeholders; increasing international openness; and preparing [staff] for the knowledge society. Given the central role of ICT in knowledge management, it is also important for organisations to take cognisance of Junnarkar and Brown's (1997) four points in the implementation of ICT as a KM enabler, namely:

- Developing enterprise-wide standards for the ICT infrastructure in order to link people to people, and people to information
- Linking IT investments to the firm's overall knowledge management strategy
- Implementing IT tools to access explicit knowledge by knowledge managers
- Establishing knowledge management partnerships that bridge information systems and human resources.

CONCLUSION

Knowledge management is as relevant to SMEs as it is to large organizations. It has successfully been applied in corporate sectors to improve productivity, profitability, customer relations, efficiency, transparency, accountability, innovation, and the overall quality of service delivery. It is therefore imperative for business enterprises and public sector organisation alike to infuse KM in their operations and strategic processes, as this would enable them to compete effectively in the global e-market environment. KM holds great potential for organisations in the knowledge economy, but in order for these benefits to be realised, it is important to understand how to align three core resources, namely people, processes and technology. Organisations that invest in new technology without understanding their organisation's and basic human patterns of information sharing are bound not to achieve their full investment potential. Leveraging an organisation's know-how and best practices can make a dramatic difference to the organisation's basic levels of competence, flexibility, responsiveness, and its customer relationships. As organisations the world over contemplate adopting KM practices, Davenport and Prusak's (1998) Principles of KM are worth noting, namely that: KM is expensive; KM needs both people and technology; KM is highly political; KM needs knowledge managers; KM must be practical and simple; the creation and sharing of knowledge is unnatural; knowledge work processes must be promoted and supported through KM; access to knowledge is only the beginning of KM; and KM never stops.

REFERENCES

Agarawal, A. M. *Knowledge management application in higher technical institutions in India*. Retrieved on September 9, 2008, from http://www.iamot.org/conference/index.php/ocs/4/paper/viewFile/982/405

Allen, P. A. (1988). *Missions of colleges and universities*. San Francisco: Jossey-Bass.

Ardimato, P., Boffoli, N., Cimitile, M., Persico, A., & Tammaro, A. (2006). *Knowledge packaging supporting risk management in software processes*. Cambridge, MA: ACTA Press.

Australian Bureau of Statistics (ABS). (2001). *Small business in Australia*. Canberra, Australia: Australian Bureau of Statistics.

Bhatt, G. D. (2001). Knowledge management in organisations: Examining the interaction between technologies, techniques, and people. Journal of Knowledge Management, 5(1), 68–75. doi:10.1108/13673270110384419

Blogie. (2007). *Knowledge management in SMEs*. Retrieved on November 18, 2008, from http://technopreneurship.wordpress.com/2007/02/11/knowledge-management-in-smes/

Bollinger, A. S., and R. D. Smith (2001). Managing organizational knowledge as a strategic asset. Journal of Knowledge Management, 5(1), 8–18. doi:10.1108/13673270110384365

Corall, S. (1999). *Knowledge management: Are we in the knowledge management business?* Retrieved on September 2, 2008, from http://www.ariadne.ac.uk/issue18/knowledge-mgt/

Davenport, T., & Prusak, L. (1998). *Working knowledge*. Boston: Harvard Business School Press.

Drucker, P. (1987). *The frontiers of management.* London: Heinemann.

Dyer, G., and N. McDonough (2001). The state of KM. Knowledge Management, 3(70), 33–38.

Grey, D. (1998). Knowledge management and information management: The differences. Retrieved on November 28, 2005, from http://www.smithweaversmith.com/km-im.htm

Handzic, M. (2006). Knowledge management in SMEs: Practical guidelines. *CACCI Journal, 1*. Retrieved on November 18, 2008, from http://www.cacci.org.tw/Journal/2006%20Vol%201/Handzic-mar2004.pdf

Hansen, M. T., N. Nohria, and T. Tierney (1999). What is your strategy for managing knowledge? Harvard Business Review, 77(2), 106–116.

Herd, J. (2001, June 7). *Taxonomy and knowledge organisation [slide presentation]. Given at the Library of Congress,* West Diningroom.

Hylton, A. (2002). *A KM initiative is unlikely to succeed without a knowledge audit.* Retrieved on September 2, 2008, from http://www.hyltonassociates.com

Hylton, A. (2003). Why the knowledge audit is in danger. Retrieved on March 30, 2007, from http://www.annHylton.com

International Records Management Trust. (2004). *The e-records readiness tool.* London: IRMT.

International Telecommunication Union. (2005). *ITU and Orbicom publish ICT opportunity index: A step towards the implementation of the WSIS' plan of action*. Retrieved on April 17, 2007, from http://www.itu.int/ITU-D/ict/publications/dd/summary.html

Kidwell, J. J., M. L. V. Karen, and S. L. Johnson (2000). Applying corporate knowledge management practices in higher education. Education Quarterly, 4, 28–33.

Kok, A. (2007). Intellectual capital management as part of knowledge management initiatives at institutions of higher learning. The Electronic Journal of Knowledge Management, 5(20), 181–192.

Leonard, A. (1999). A viable system model: Consideration of knowledge management. *Journal of Knowledge Management Practice*.

Maponya, P. M. (2004). *Knowledge management practices in academic libraries: A case study of the University of Natal, Pietermaritzburg Libraries*. Durban: University of KwaZulu Natal.

McLean, D. (1999). Knowledge works. Knowledge IQ Management, 1(3), 20–24.

Microsoft. (2000). *Digital dashboard: Business process assessment guide*. Retrieved on September 2, 2008, from http://www.destinationcrm.com/km_km_article.asp?id=24

Microsoft. (2001). *Introducing the dashboard site for Microsoft sharepoint portal server 2001*. Retrieved on September 2, 2008, from http://www.microsoft.com/sharepoint/techinfo/digitaldash.asp

Milam, J. H. (2001). *Knowledge management for higher education*. Virginia: University of Virginia.

Nonaka, I., & Takeuchi, H. (1995). *The knowledge creating company*. London: Oxford University Press.

O'Really, T. (2005). *Web 2.0: Compact definition*. Retrieved on September 23, 2007, from http://radar.oreilly.com/archives/2005/10/web_20_compact_definition.html

Oosterlinck, A. (1999). *Knowledge management in post-secondary education: Universities*. Retrieved on September 8, 2008, from http://www.oecd.org/dataoecd/46/21/2074921.pdf

Organisation of Economic Cooperation and Development. (2003). *The learning government: Introduction and draft results of a survey of the knowledge management practices in ministry/departments/agencies of central government*. Retrieved on November 29, 2005, from http://www.oecd.org/dataoecd/54/2/33688562.doc

Porter, M. E. (1985). *Competitive advantage: Creating and sustaining superior performance*. New York: Free Press.

Rasheed, N. *The impact of knowledge management on SMEs*. Retrieved on November 18, 2008, from http://www.knowledgeboard.com/download/2539/THE-IMPACT-OF-KM-ON-SMEs.pdf

Skyrme, D., & Associates. (2002). *KM benefits tree*. Retrieved on November 29, 2005, from http://www.skyrme.com/tools/bentree.htm

Srinivas, H. (2007). *Packaging knowledge: An information continuum*. Retrieved on May 20, 2007, from http://www.gdrc.org/icts/continuum.html

Storey, J., & Quintas, P. (2001). Knowledge management and HRM. In J. Storey (Ed.), *Human resource management: A critical text*. London: Thomson Learning.

Sveiby, K. E. (2001). *What is knowledge management?* Retrieved on March 30, 2007, from http://www.sveiby.com/Portals/0/articles/KnowledgeManagement.html

United Nations. (2008). *UN e-government survey 2008: From e-government to connected governance*. New York: United Nations.

Wagner, R. (2008, August 21-22). *The power of knowledge*. Paper presented at the Libraries and Information Management Forum, Sandton Hilton, Johannesburg, South Africa.

White, T. (2004). *Knowledge management in an academic library*. Retrieved on June 18, 2006, from http://eprints.ouls.ox.ac.uk/archive/00000815/01/Tatiana_White_KM_article.pdf

Wimmer, M. A. (Ed.). (2004, May 17-19). *Fifth working conference on knowledge management in electronic government (KMGov2004)*, Krems, Austria. Retrieved on November 21, 2005, from http://falcon.ifs.uni-linz.ac.at/kmgov2004/

Wimmer, S., & Wolter, H. (2000). *Situationen und Perspektiven des indutriellen Mittelstandes in der Bundesrepublik Deutschland*. Retrieved on November 18, 2008, from http://www.ifm-bonn.de.ergebnis/77nf.htm as cited by Wickert and Herschel 2000:329.

Chapter 13
E-Records Management and Freedom of Information for SMEs

INTRODUCTION

While records management in general and e-records management in particular have received serious attention in large enterprises, the same cannot be said for SMEs. This is despite the emergence of an economy where transactions are increasingly carried out electronically, and where the management of e-records emanating thereof must be meticulously managed to obviate disaster. Iron Mount Digital (2007), a leading provider of data backup/recovery and archiving software based in the UK, warns in its findings (based on industry research) that SMEs are reportedly failing to put in place disaster recovery plans in the belief that such recovery plans are the domain of larger companies. Such SMEs could face total shut down or bankruptcy unless they developed contingency plans to ensure that normal operations are resumed follow-

DOI: 10.4018/978-1-60566-420-0.ch013

ing a serious incident. In these plans, SMEs should include a disaster contingency team, assessing key areas of risk across the business, planning for recovery, and reviewing and communicating the plan to employees. It is important for SMEs to continuously backup crucial company information so that in the event of a disaster, the company can be safe in the knowledge that it can restore company data.

Generally, records are strategic and operational assets that are vital to the operation of the state. Like traditional paper-based records, e-records support the day-to-day operation of government services and interactions with citizens and private and public sector partners. As government services move online, e-records become the basis for confirming pensions and other entitlements; registering births and deaths; verifying citizenship; certifying voting rights; enabling the collection of taxes and census enumeration, supporting financial management; enabling audits and evaluations; supporting litigation; and documenting inter-governmental agreements, to name a few (IRMT, 2004). Both paper-based and electronic records are valuable assets that need to be well managed and protected. In addition to providing essential evidence on organisational activities, transactions and decisions, records also support business functions and are critical for the assessment of organisational performance. Without reliable records, governments cannot manage state resources or civil service, or provide services such as education and healthcare effectively. And without effective systems to manage them, governments cannot be held accountable for their decisions and actions, meaning that the rights and entitlements of citizens and corporate bodies cannot be upheld.

The benefits of well managed records in both public and corporate organisations of all sizes include (Kemoni, 2008):

- The ability to mitigate the considerable risks associated with poor records management practices, specifically through accountability, transparency, sound corporate governance and public sector efficiency
- Enabling compliance with statutory requirements
- The ability to provide enterprise-wide access to documents, records and information resources contained within multiple databases
- The ability to use electronic documents and records as inviolate and credible evidence
- Instilling knowledge of fundamental records management practices and how they relate to FOI (Freedom of Information) and information and privacy principles
- Increasing productivity and individual accountability.

Well managed records contribute to the smooth operation of an organisation's programmes by delivering services in a consistent and equitable manner. Through proper management, accurate, timely, relevant and credible information can be made available to the organisation. Business efficiency largely depends on the information held in records; consequently, good documentation is necessary for efficient planning, financial analysis, transaction processing, product development, quality control, marketing, customer services and other business operations. Records are also essential for the effective functioning of private and public organisations because they document the decisions and activities of governments and organisations and serve as a benchmark by which future activities and decisions can be measured. Furthermore, without good records, members of staff are forced to take decisions on an ad hoc basis without the benefit of institutional memory (IRMT, 1999: Cashmore and Lyall, 1991; American Records Management Association, 2004; Lawoye, 1986; The World Bank, 2006; Robeket al., 1996).

Good records management is important to ensure that records of archival and historical value are preserved, and those which are no longer in use are destroyed. Records management practices also provide a clear accountability structure and audit trails, and enhance good governance. As more citizen-state interactions occur in electronic form, citizens will expect that their rights are as well protected and documented in an electronic environment as they would be in a paper-based one. This can only be achieved if the records generated through e-government are carefully managed with the help of systems that provide constant intellectual and physical control. The aim must be to preserve the combination of content, context and structure, which give electronic records meaning over time; to protect the fragile media from degradation; and to ensure efficient access.

RECORDS MANAGEMENT, SMEs AND FOI

The SME sector makes significant contribution to public sector services. For example in the UK, SMEs are a critical part of the economy, accounting for 99% of all businesses in the country, and over half of the UK's employment and turnover. The EU's Lisbon Agenda has set the target of making the EU the most dynamic and competitive knowledge-based economy by 2010, and the manner in which businesses access and engage with communications technology is a fundamental part of achieving this goal (Curie, 2006). Consequently, FOI applies as much to large companies as it does to SMEs in the UK, and the contribution of SMEs is included in the final (annual) assessment of the size of the public services industry and its contribution to the UK economy (McGuire, 2008). The FOI legislation is premised on the principle that effective records management enables authorities to enforce wider government agendas, such as increased openness, transparency, trust and accountability in the public sector (Shepherd and Stevenson, 2008). FOI enhances transparency in the government's procurement of services, especially when it comes to opportunities for SMEs to work on government contracts. Better governance of the procurement process can enhance and create more opportunities for SMEs.

Within the House of Commons, there is a strong argument in support of the notion that information should be shared closely and that bridges should be built with SMEs in order to promote openness and transparency. In turn, this would encourage the greater involvement of SMEs in tendering for government contracts. The generally held belief here is that while governments use a huge number of suppliers, their contracts are closed to the vast majority of SMEs. Communicating with suppliers and managing relations with them remain critical issues – how do you communicate fairly, openly and in some detail with over 30,000 different organisations? (CSR Case Studies, 2004).

FOI is increasingly being enacted by many countries, particularly in the developed world, although a few developing countries are following suit. By 2006, there were over 30 countries worldwide that had enacted FOI legislation; some had passed the legislation and were still awaiting its implementation, while others were contemplating its enactment (Sebina, 2006). The UK's 2000 Freedom of Information Act came into force in January, 2005. The Act imposes significant duties and responsibilities on public authorities to provide access to information. The public authorities have to know what information they hold and manage so that they can retrieve information effectively, deal expeditiously (within 20 days) with FOI requests, and disseminate information through a publication scheme.

The UK FOI Act categorically states that any person making a request for information from a public authority is entitled to be informed in writing by the public authority on whether it holds information fitting the description specified in the request, and if so, to have that information communicated to the person in question. However, access to such publicly held information is subject to the proviso that when a public authority reasonably requires further information in order to identify and locate the requested information and has consequently informed the applicant of that requirement, the authority is not obliged to comply unless or until it is supplied with the additional sought after information. The information with which the applicant is to be informed, or which is to be communicated, is the information held at the time of the request; hence any amendment by the applicant to the initial request for information made between that time and the time before the response to the request was sent may be ignored. A public authority can be taken to have complied with a request for information if it has sent the requested information to the requesting party (Office of the Public Sector Information, 2000). The individual or entity requesting information has to comply with the following proviso: the request must be put in writing; the applicant must state their name and address; and the applicant must describe the information they seek.

Scotland has a separate FOI from that of the UK. The Freedom of Information Act of 2002 was passed by the Scottish Parliament on the 24th of April of that year, although it only came into full force on the 1st of January, 2005. The Act aims to promote openness and accountability among Scottish Public Authorities by introducing a general statutory right of access to the various types of recorded information held by Scottish public authorities (Burness, 2004).

In the US, the Federal Bureau of Investigation defines what kinds of information can be requested under FOI. This information includes: information about an organization, business, investigation, historical event, or incident; information about a third party; information about a deceased person; and information about one's self. The Record Information Dissemination Section (RIDS) of the FBI is responsible for effectively planning, developing, directing and managing responses to requests for access to FBI records and information. The requests and disclosure must comply with the Freedom of Information and Privacy Act. RIDS efforts are directed to appropriately release information in an efficient and effective manner while protecting legitimate law enforcement, foreign policy, and national security and defence interests, and remaining forthcoming to the American public (US Department of Justice: FBI, 2008).

The African Union, through its Charter on Human and Peoples' Rights, recognizes the importance of access to information. The Charter recognizes that access to information is part of freedom of expression. It also states that freedom of expression, including the right to seek, receive and share information and ideas, whether orally, in writing, in print, in the form of art, or through any other form of communication (and across frontiers), is a fundamental and inalienable human right and an indispensable component of democracy (Sebina, 2006).

FOI and sound records management are inextricably linked. Records have a unique character that imposes special responsibilities on the agencies that preserve and manage them. State records not only document past decisions, but also establish and protect the contemporary rights and responsibilities of both the government and the governed. Sebina (2006), citing John Bolton - a Records Officer working for the British Columbia Ministry of Skills – argues that without a substantial and comprehensive records management programme in place, the FOI legislation would not be worth the paper it was written on. Records could neither be identified as existing, nor obtained from their storage. The place of records in achieving the objectives of FOI is stressed in a draft Code of Practice on the Management of Records Under Section 46 of the UK's Freedom of Information Act (2000), where it is stated that: "Any freedom of information legislation is only as good as the quality of the records to which it provides access. Such rights are of little use if reliable records are not created in the first place, if they cannot be found when needed, or if arrangements for their eventual archiving, or destruction are inadequate" (Department of Constitutional Affairs, 2002). For records to meet the objectives of the FOI legislation, they must be accurate, complete, timely, appropriate for retention, relevant, adequate, credible, engaging, readily accessible, and likely to be needed in the future.

Sebina (2006) explains that the Freedom of Information (FOI) legislation is mostly adopted on the premise that good records management exists. However, it is pertinent that the functionality of records management in creating, managing and making records available for access internally within government and externally to citizens, is established while the legislation is being planned for. Through the planning process, the capacities of records management in providing access to information will be known. Access to information is an important process of democratic governance because through it, democratic states are better able to ensure that citizens can gain access to official information. At the same time, the laws that govern access to information ensure that the privacy of citizens is protected by restricting access to their personal information. These laws also act as a defence against attempts to access information that may compromise the integrity and security of the state.

The importance of FOI to enhance access to publicly held information cannot be over emphasised. Governments, by their mandate, hold in their custody various types of information, such as international accords; negotiating briefs; policy statements; minutes of discussions with investors, donors and debtors; cabinet deliberations and decisions; parliamentary papers; judicial proceedings; details of government functioning and structure; intra-governmental memos; executive orders; budget estimates and accounts; evaluations of public expenditure; expert advice; recommendations and guidelines; transcripts of departmental meetings; statistical data; reports of task forces, commissions and working groups; social surveys and analyses of health, education and food availability; assessments of demographic and employment trends; analysis of defence preparedness and purchases; maps; studies on natural resource locations and availability; proof of the quality of the environment, water and air; and detailed personal records (Sebina, 2006). Access to public information in the custody of the state under FOI legislation is guided by principles of maximum disclosure; limited scope of exemptions; proactive disclosure of information; implementation and monitoring; access to information fees; and the precedence of disclosure (Sebina, 2006).

The principle of maximum disclosure intimates that access to government information should be viewed as a right rather than a privilege, and posits that all government information should be accessible unless justifiable exemptions exist. The limited scope of exemptions principle states that in limiting the disclosure of information, governments are obliged to state reasons working against the disclosure, and these should be derived from the legislation. Proactive disclosure of information requires the agencies responsible for public information to release certain aspects of information proactively without waiting for requests to be made for their release. The principle of implementation and monitoring states that the success of any FOI legislation is dependent on effective enforcement and monitoring. Laws should specify how requests made for access to information are to be assessed or decided on, and what remedial provisions exist. With regard to the principle of access to information fees, FOI recognises that satisfying requests for access to information might stretch the already minimal resources that governments have. As a result, this principle affirms that governments may levy fees for requests and for retrieving and providing the requested information, whether photocopied, digitised or in any other format. Finally, the principle of the precedence of disclosure stipulates, among other things, that all the laws of the country should be consistent with the provisions of the FOI legislation. Those that are not should be repealed or amended to align them with the canon of FOI (Sebina, 2006).

E-RECORDS READINESS

By and large, crucial company information and knowledge is embodied in the records of business transactions. The subject of e-records management therefore applies equally to both large enterprises and SMEs. The globalization of trade (particularly in the case of electronic commerce) and the move by governments towards e-governance in an effort to cut down on the cost of bureaucratic processes and reduce the time spent on paper-based records, is generating large amounts of e-records. In order for any organisation to transact business in such an environment and effectively manage the resulting outpour of e-records, the organisation needs to have attained a certain measure of e-readiness, or e-records readiness so to speak. E-readiness, as already defined, refers to a community that provides high-speed, constant access to ICTs and their applications in various sectors in a competitive market, including government (Bridges.org, 2001). In contrast, e-records readiness refers to the capacity to create, manage, share and use electronic information (and related technology) to improve governance and sustain international trade and innovation; improve global security; and support other related activities in an increasingly interconnected world (Lipchak and Donald, 2003:2).

The E-records Readiness tool - developed by the London-based International Records Management Trust (IRMT) in partnership with public sector institutions, international donors, professional associations and academic institutions - is used for assessing e-records readiness. The E-records Readiness Tool was developed to enable governments to conduct high-level assessments of key areas of e-records readiness in relation to other aspects of e-government, and determine whether their records and information management infrastructure is capable of supporting e-government initiatives (IRMT, 2004). The tool uses a brief questionnaire to provide a risk assessment of e-records readiness in government at national and enterprise levels. The areas addressed by the tool are varied, and include: staff competencies in maintaining software and hardware; training programmes for information management staff; human resource strategies for e-records; telecommunications infrastructure to support the growing volume of work; adequacy of the electrical power supply; formal records management practices; information management policies & responsibilities; and tools and procedures for information management. Other aspects covered by the tool include:

- Information management products and technologies
- Internal and public awareness programmes for information management
- Compliance with information management procedures (i.e. security, backup, and confidentiality)
- Documentation standards and system engineering procedures for ICT
- Guidelines for the management of electronic records
- National ICT strategies
- Supportive legal and regulatory frameworks for information management
- Freedom of information and protection of privacy.

E-RECORDS GENERATION AND USE

For a long time the corporate sector has been a front runner in technology adoption, but now there is increasing implementation of ICT in government driven by public sector reforms. Public sector reforms refer to interventions that affect organisations' performance and the working conditions of employees in central, provincial or state government (DFID and University of Birmingham, 2003). Public sector reform initiatives first started being implemented by the World Bank in developing economies in the 1980's, and initially focused on down-sizing and controlling salary costs, primarily through job reduction, performance assessment, monitoring, transparency, benchmarking, decentralisation, regulation, and sound financial management. These reforms coincided with the ICT revolution, specifically of the Internet and the World Wide Web, which presented governments with an opportunity to enhance efficiency, accountability and transparency in the management of public affairs. Shepherd and Yeo (2003) explain that records management using technological solutions is almost certain to form part of any systems design process of an organisation. These technological solutions may include adding enhanced records management functionality to existing software and/or acquiring new applications.

Public sector reforms, coupled with the increased use of ICT in corporate and government organizations, have accelerated the generation of e-records and the need for the proper management of such records. The ISO 15489:2001 defines a record as information that is created, received and maintained as evidence by an organization or person in pursuance of legal obligations or in the process of business transactions. By contrast, an e-record refers to recorded information, documents or data that provide evidence of the policies, transactions and activities carried out in e-government and e-commerce environments (IRMT, 2004).

E-records may be categorized as text files (files produced using word processing programmes or other software); data files (computer files that store numeric and sometimes textual information as quantitative values so that numbers can be manipulated using arithmetic processes); analogue audio and visual records (sound documents and images that are replayed at a later time); disaggregated data (information collected through remote sensing systems); databases (structured collection of interrelated data); machine instruction sets (records created through the action of intelligent machines); image files (records containing computer images that generally exist in hard copy format before being converted into images) and digital documents (files consisting of numeric data, images, or sound recorded digitally in one uniform structure). Specific e-records may include e-mail, Internet content, documents, spreadsheets, drawings, etc.

E-RECORDS MANAGEMENT

Records management is defined as the field of management responsible for the efficient and systematic control of the creation, receipt, maintenance, use and disposition of records, including the processes for capturing and maintaining evidence of and information about business activities and transactions in the form of records (Mutula and Wamukoya, 2007.

Sound records management involves the following (ISO 15489, 2001):

- Creating, approving, and enforcing records policies, including classification systems and a records retention policy
- Developing a records storage plan, such as the short and long term housing of physical records and digital information
- Identifying existing and newly created records, classifying them, and storing them according to standard procedures

- Coordinating the access and circulation of records within and even outside an organization
- Executing a retention policy to archive and destroy records according to operational needs, procedures and regulations.

The management of records is achieved using tools such as guidelines, policies, regulations, legislations and standards, from the creation stage to the stage of preservation and use. For example, the ISO 15489 standard is designed to provide a framework for planning and implementing a records management program and maintaining structured records systems. ISO 15489 is comprehensive in its cover of current and past records, and provides a clear categorisation of its requirements. This also makes it an obvious choice on which to base records management programmes.

ISO 15489:2001 (2001) states that sound records management practices should include:

- Setting policies and standards
- Assigning responsibilities and authorities
- Establishing and promulgating policies and guidelines
- Providing a range of services relating to the management and use of records
- Designing, implementing and administering specialized systems for managing records
- Integrating records management into business systems.

E-records management is yet to receive the attention that it deserves. The E-discussion on Evidence-based Governance in the Electronic Age, a global forum held in Johannesburg, South Africa, from the 27th – 31st of January 2003, highlighted the need for collaborative action in managing electronic records within electronic government requirements. Although it was clear that all the countries involved were moving rapidly into an e-working environment, participants from low resource environments complained about the dearth of investments and solutions for managing electronic records. The discussion served as a virtual audit, revealing that while nearly all commonwealth governments had ambitious strategies for ICT development and e-government, very few had introduced policies, standards, systems and procedures for managing the electronic records that these applications were likely to generate (IRMT, 2003).

E-RECORDS MANAGEMENT MODELS

There are various records management models, including: the Records Life-cycle Concept/Theory; the Records Continuum Model; the International Council on Archives (ICA); Electronic Records Management Guidelines (Model); the National Archives of Australia Records Management Guidelines, AS ISO 15489:2002; the National Archives of Australia Digital Recordkeeping Guidelines, 2004; the National Archives and Records Service of South Africa Guidelines (Model); and The Victorian Electronic Records Strategy Model [VERS] (Tough and Moss, 2006). However, these models have largely originated from two commonly used approaches, namely the Records Life-Cycle and Records Continuum approaches. A few weaknesses in the Records Life-Cycle concept led to the development of the Continuum model. For example, the Records Life-Cycle model was criticized because (Yusof and Chell, 2000:137) it was not developed for the management of electronic records. Moreover, with changes in technology, the record was prone to transformation and conversion, but the model failed to take these changes into consideration.

The Continuum model is preferred over the Records Life Cycle model because it caters for both electronic and paper-based records. The Continuum model is divided into four dimensions, namely: create, capture, organize and pluralize (Upward, 1998). The dimensions do not have boundaries, but are instead coordinated and integrated. This means that different record keeping activities can happen simultaneously across the four dimensions. The first dimension encompasses the actors who carry out the act (decisions, communications, acts, etc), the documents that record the acts, and the trace and representation of the acts. The second dimension consists of personal and corporate recordkeeping systems that capture documents, while the third dimension encompasses the organisation of recordkeeping processes. This dimension is concerned with the manner in which a corporate body or individual define their recordkeeping agenda, and in so doing, can be viewed as the archive or memory of business. The fourth dimension concerns the manner in which the archives are coalesced into an ambient framework in order to provide a collective social, historical and cultural memory of the institutionalised social purposes and roles of individuals and corporate bodies. Therefore, the concept of "records" in the Continuum model is inclusive of records of continuing value (archives). Moreover, the Continuum model provides a way to conceptualize recordkeeping in organizations and does the following (McKemmish, 1998):

- Identifies key evidential, recordkeeping and contextual features of the continuum and arranges them in relation to each other
- Represents the multidimensional nature of the recordkeeping function

- Maps the evidential, recordkeeping and contextual features of the continuum against the dimensions of the recordkeeping function
- Positions recordkeeping in a broader socio-legal and technological environment.

The actions under the records Continuum model (Millar, 1997:14) include the following:

- Identification and acquisition - records management actions are the creation or acquisition of records, while archives management actions relate to the selection and acquisition of archives
- Intellectual control - records management actions include the classification of records within a logical system, while archives management actions relate to the arrangement and description of archives
- Access - records management actions relate to the maintenance and use of records, while archives management actions relate to the description of archives
- Physical control - records management actions enable disposal through the destruction of records or their transfer to national archives, while archives management actions relate to the preservation of archives.

There are other records management models that can be applied to electronic records. For example, the ICA Committee on Electronic Records model was designed to help archival institutions reposition themselves in order to address the management of archival electronic records. The model discusses the technological, organizational and legal trends that impact on the ability of organizations, including archives, to keep and manage records that are in electronic form. The key issues addressed by the model are: records in a database environment; records and archives in the electronic age; strategies for managing electronic archives; preservation of electronic archives; and access and legal and policy implications of electronic archives (Kemoni, 2008).

The National Archives of Australia (2004) is another records management model that was developed to cater for digital recordkeeping. It provides guidelines that address the creation, management and preservation of digital records. The guidelines cover the following areas: importance of digital records, digital recordkeeping frameworks, creating digital records, creating information about digital records, determining how long to keep digital records, and storing and securing digital records. It also covers business continuity planning for digital records, and the preservation, access and disposal of digital records. The guidelines also provide information on how to manage common types of digital records such as electronic messages, web-based digital records, records subject to online security processes, and records in business information systems (Kemoni, 2008). The major strength of the National Archives of Australia Digital Recordkeeping Guidelines (2004) is that they are comprehensive in assisting organizations on the topic of e-records management. The guidelines also address the key challenges associated with the management of electronic records, including the creation and capture of electronic records, storage, security, appraisal, disposal, and business continuity planning for electronic records. Furthermore, the guidelines provide help relating to the management of web-based digital records.

The VERS (Victorian Electronic Records System) Model (2004), developed in Australia, is a combination of standards, guidelines and implementation projects that can be used to archive electronic records created or managed by the Victorian government. The model (Kemoni, 2008):

- Specifies a single, minimal framework for the management of electronic records
- Specifies a long-term format for the capture of e-records that need to be preserved over an extended period
- Works in conjunction with an agency's existing recordkeeping and business practices.
- Ensures that all records are stored for the long-term to facilitate the viewing of records in the distant future, regardless of the system that created them
- Specifies methods to automate the capture of records from desktop and agency business systems
- Ensures that records in the future will be understood in context
- Details methods for securing electronic records so that unauthorised changes can be detected.

The University of Pittsburgh's Electronic Records Management Model (2004), cited in Kemoni (2008), is integrated and built upon three existing branches of knowledge, namely business process improvement, information systems development, and electronic records management and archival requirements. The model was developed to integrate records, archives' management requirements and systems development methodologies in an electronic records management environment. The model is more suitable for records management studies dealing with the design and implementation of electronic records management systems.

E-RECORDS AND INFORMATION SECURITY IN AN E-ENVIRONMENT

As more businesses go online, concerns about security continue to grow, particularly because the Internet has become a common platform for online fraud (e.g. credit card theft). Developing online security requires policies to be put in place to protect businesses that are dependent on technology.

The need for appropriate policies to mitigate security risks is even more urgent given the rapid spread of the Internet across governments and businesses. The expansion of computers in the global economy, their quick assimilation into households, and new developments relating to computer networks have created even more information security concerns (Ministerstvo, 2002). These developments raise several questions, for instance, can the provenance of electronic messages be accurately determined? What infrastructure is necessary for the secure transmission of electronic information? How can we guarantee that the contents of a message have not been altered during transmission? (MacNeil, 1998).

The security of data transmission depends on the encryption of data and the authentication of the source and destination of the data, which are provided through public key security systems. The primary obstacle to the widespread deployment of such systems is the lack of national support structures, such as public key infrastructure (PKI). PKI associates authentic public keys with authorized users; administers the management of the keys, including their generation, distribution, and deletion; and stores, retrieves, and archives keys (Jutla et al., 2002). The goal of companies that maintain PKI is to validate each party involved in an Internet transaction; that is, they guarantee and confirm that the individuals exchanging information online are who they claim to be. Government legislation that equates electronic signatures to handwritten signatures is another way to build the public's confidence in the security of Internet transactions and promote the use of electronic communications.

Because most business enterprises deal with local, national and international banks continuously, mechanisms to make it easier to transact B2B commerce between SMEs in different countries through financial sector networks should be put in place. For example, *Identrus* is a PKI-type application that permits financial institutions to issue digital certificates to their business customers, and allows the business customer to seamlessly deal with multiple financial institutions. In other words, a single digital certificate can enable the customer to settle transactions with many different institutions. *Identrus* applications include e-billing, secure messaging, and document handling. The Bank of America, Chase, Citibank, Wells Fargo, ABN Amro, and the Deutche Bank are among the institutions enabling secure commerce through *Identrus* (Jutla et al., 2002). The implications of *Identrus* for SMEs are in the area of facilitating international trade. If, for example, an SME in one country wishes to do business with an SME in another, and if both businesses are using banks on the *Identrus* network, there are intermediaries between the two that implicitly state that they know about these businesses, thereby instilling trust.

Another example of a security application is *TrustInfo*. *TrustInfo* is the trust seal used by the International Chambers of Commerce, the World Chamber of Commerce, and the Paris Chamber of Commerce. The rationale behind the application of this technology is that in order for businesses to obtain the *TrustInfo* seal, the business must be a member of a local chamber of commerce, and therefore be known to a trusted third party. The accreditation process for the seal requires four parties: the business registry in the local area, the local chamber, a contact person in the SME Company, and a local bank. *TrustInfo* is used in B2B transactions between small-sized enterprises in countries such as France, Ireland, Canada, the US, the UK, Greece, Romania, Bulgaria, and many of the Mediterranean and South East Asia countries. The chambers of commerce promote online alternative dispute resolution for their SME members.

Authentication is another way to enhance the security of data. Authentication refers to the provision of proof that the person or entity requesting a particular service is the one authorized to use the service. Authentication involves identifying the originator of the information, while non-repudiation of electronic messages ensures that neither sender nor the recipient can subsequently deny that they were involved in the transaction.

Generally, when a record is said to be authentic, it means that it is real or genuine (Collins Pocket English Dictionary, 1996). A cursory glance at a thesaurus demonstrates the varied range of meanings that authenticity has evoked in the modern era. Fine distinctions include *authoritativeness* (i.e. assertiveness, commanding, definitive, official, sound, sovereign), *certainty* (i.e. assurance, confidence, faith, fact), *purity* (i.e. brilliance, blamelessness, chastity, wholesomeness), and *validity* (i.e. authority, cogency, substance) [Collins Paperback English Thesaurus, 1995]. In the legal context, authenticity is referred to as *authenticum* (an original instrument or writing) [Black, 1995] and requires the attestation of a public authority to guarantee that the document is real. In the world of computer technology, authentication is seen as the process of affirming that remote people or things are who or what they claim to be (Ford and Baum, 1997).

There are various kinds of e-authentication, some of which are described below:

- **Digital signatures:** a digital signature refers to an electronic method of identification that is executed or adopted by a person with the intent to be bound to a record or to authenticate it. An electronic (or digital) signature is usually unique to the person using it and is therefore capable of reliable verification, and can be linked to a record such that if the record is changed, the electronic signature is invalidated. It was not until the 60s and 70s, when electronic data interchange began to support inter-business [or business-to-business] (Keen and Balance, 1997) trading, that electronic authentication became more common. From the 90s into the 2000s, and increasingly today, various federal and national governments have been urging citizens to consider conducting business electronically. For example, in 2000, the governments of the United States (Internal Revenue Service, 2000), Canada and various European countries were encouraging their revenue and tax departments to increasingly conduct their business over the Net. In these business-to-consumer transactions, the reliance on the digital signature as a means of ensuring authenticity has grown increasingly prominent. At the international level, digital signatures are being used to sign multilateral contractual agreements.

Despite the increasing use of digital signatures in e-environments, they are perceived not to serve the same function as handwritten signatures. The United Nations has been at the forefront in advocating for the transference of the functional equivalence of handwritten signatures to digital signatures. In its model law, the United Nations Commission on International Trade (1997) outlined the possible range of functions traditionally performed by handwritten signatures. Key among them, are that they are used to:

- Identify a person
- Provide certainty of the personal involvement of that person in the act of signing
- Associate that person with the content of a document
- Attest to the intent of a party to be bound to the content of a signed contract
- Attest to the intent of a person to endorse the authorship of a text
- Attest to the intent of a person to associate himself/herself with the content of a document written by someone else
- Attest to the fact that a person had been at a given place
- Attest to the time when a person had been at a given place.
    - **Digital certificates:** These are security devices that consist of an electronic document (similar to a digital signature) that is usually transmitted through the Internet. The certificate is basically an assurance that the information is indeed from the source that it purports to come from, and has not been interfered with in any way. A Certificate Authority is an

agency that manages the distribution of digital certificates or signatures (InstallShield DevStudio, 2003).

E-RECORDS MANAGEMENT CHALLENGES

Electronic records and the systems that support e-records are complex and fragile. They are created and maintained using technology platforms and standards that change frequently; are stored on media that deteriorate over time; and are often supported by weak accountability and management frameworks and practices. E-records require a combination of software and operating systems to enable access to information they contain. With rapid evolution of technologies, this creates problems of technological obsolescence where for example older technologies are not compatible with newer ones. Moreover, the staff competencies, skills and tools needed to manage e-business processes and e-information in a shared work environment has not been adequately developed in many public sector organizations, especially in developing countries. At policy level, senior officials and legislators are often unaware of the need to manage electronic records over time. At the planning and operational level, systems designers and IT specialists tend to focus primarily on current information needs without paying attention to long-term preservation requirements. Finally, among records and information managers and national archivists, especially in developing countries, there is insufficient capacity and training to articulate e-records issues and provide guidance and input to policy makers (IRMT, 2003b).

The pace of technological change and the highly flexible environment in which electronic records are generated and stored makes them more volatile and vulnerable than paper-based records. Managing e-records raises specific issues peculiar to this format. It is more difficult to ensure that the content, context and structure of records is preserved and protected when the records do not have a physical existence. The current technological environment tends to generate large volumes of records, and also allows records to be easily altered and manipulated. This creates significant challenges in maintaining adequate evidence of organisations' business operations. Fragility of e-records media, and voluminous accumulations of e-mails are examples of some of the technological problems that organisations experience with managing e-records (The Department for Administrative and Information Services, 2002; An, 2001; American Records Management Association, 2004).

In a study to assess the technological infrastructure and needs in East and Southern Africa, Mutiti (2001) revealed that limited progress had been made in the area of managing the electronic records created by public institutions. Agencies with the statutory responsibility for records in the region have been working to address computerization issues, but most of this work has focused on automating the description of their holdings. Specifically, the findings revealed that the most common application of computers in these agencies was word processing; most countries had no specific legal or administrative frameworks from which to operate an electronic records programme; and most automation programmes were in the initial stages of development. The study identified various problems impeding the automation of archives and the management of electronic records, including the lack of awareness among records officers about their role in national e-records management programmes; lack of prioritization of automation functions and services; an inability to select appropriate hardware and software; and an inability to provide guidance on electronic records management to government agencies.

Other problems relating to the management of e-records in East and Southern Africa that still bedevil the region include (Mutiti, 2001): a lack of policies and guidelines on the management of electronic records; lack of standards, practices and procedures for the management of electronic records; lack of knowledge, competencies and skills in the management of electronic records; and the mismanagement and loss of electronic records created in public institutions. The study made the following recommendations: archivists should be trained in electronic recordkeeping to enable them to play an active role in the management of electronic records; the sharing of ideas, experiences and expertise on e-records management-related issues through seminars and workshops should be encouraged; and agencies with the statutory responsibility for records should incorporate e-records management into their core functions.

POLICY AND MANAGERIAL ROLES

Chronic weaknesses in governments' record-keeping practices can adversely affect private sector investment. For example, overseas firms may hesitate to invest in a country if they feel that its courts do not handle civil cases (especially commercial cases) efficiently. Likewise, large-scale infrastructure investments, such as the construction of gas pipelines, may be delayed or may incur significant additional costs if government land registries cannot provide complete and definitive statements of titles to property. More generally, poor record keeping can contribute to a lowering of the general standard of services offered to businesses. For example, there may be delays in replies to written inquiries about the registration of businesses, licensing issues, and other matters necessary for companies to pursue their business (Mutula and Wamukoya, 2007).

Within the e-environment, the role and participation of the private sector is critical, especially with regard to e-commerce and e-business transactions. Governments must create a comfortable environment by establishing legislations and regulatory frameworks that make it easier to manage e-records. As more and more private sector and government operations are carried out online, such legislations and regulatory frameworks will be critical in ensuring that reliable evidence of all the activities and transactions are on hand to protect the rights, obligations and entitlements of all the parties involved. IRMT (2004) stresses this point, having observed that under existing legislations, courts around the world have struggled to apply the traditional rules of evidence to e-records, with the net result being inconsistencies. Thus, in order to facilitate dispute resolution and avoidance, governments need to adopt laws that establish ground rules for e-transactions, e-commerce and the use of e-signatures. Moreover, national governments need to modernise, clarify and harmonise e-commerce laws and regulations so that public and private sectors alike can make the best possible technical decisions about how to transfer e-records across jurisdictions without having to worry about how their legal rights will be affected.

In order to ensure that digital records remain accessible, strategies that involve migration, emulation and the conversion of digital records are regularly required for business and administrative purposes. It is also important to ensure the long term viability of electronic records by documenting and retaining them in hard copy (Lin et al., 2003).

It is also necessary to constantly update and improve national ICT infrastructure and policy, legislative and regulatory frameworks, and revise the capacity to implement e-records management plans. Furthermore, policies, strategies, systems and procedures are required to ensure that the e-records generated by computer systems and the metadata that gives them their authenticity are captured and preserved as reliable and verifiable documentary evidence. E-records management should be incorporated as a fundamental component of ICT and e-government strategies. Governments need to embrace innovative initiatives, such as public sector reforms that aim to re-examine records management practices, to ensure efficiency and productivity in the provision of government services to citizens.

The general strategy should aim to develop a cadre of records management professionals with the capacity to leverage records management functions for their respective countries and institutions. It should also contain various training initiatives and skills through which appropriate people would be trained to manage e-records. In addition to human resource capacity, the strategy would help by identifying the infrastructural developments and capacities that need to be undertaken. Such a strategy would determine the training needs, competencies, and skills of records personnel and other stakeholders with regard to electronic records management; define e-records and identify key issues associated with their creation, management, and use within governments; develop an e-records capacity building strategic plan for national archival institutions, records creators and other stakeholders; identify models of excellence in records and archival management operations to serve as guides for technology and skills transfer; put an implementation plan in place for e-records capacity building; and determine the resource requirements for the implementation of the projects under the strategic plan.

The intended benefits of e-records may be compromised in instances where e-government services are delivered using new ICTs unless there is adequate, dynamic infrastructure, and/or a backup plan for managing the new arrivals. Furthermore, it is important within an e-government environment for a central agency, with the statutory responsibility for records, to be either established, or called upon to coordinate e-records management activities. This agency would be tasked with ensuring that government-wide standards and practices are developed and implemented, and that appropriate facilities and resources are availed for the management of official records in digital and other formats. In order for the agency to fulfil its mandate effectively, these additional roles and responsibilities should be recognized and embedded in the respective records management acts in different jurisdictions (IRMT, 2004).

CONCLUSION

As more of their transactions get automated through e-government, leading to the generation of large amounts of records in electronic format, governments have been pushed to make the management of e-records a top priority. E-records management, e-governance and electronic business transactions are closely and inextricably intertwined. In the case of e-governance, sound e-records management enhances transparency, accountability and good governance. This in turn creates an environment that is conducive for businesses to thrive, especially SMEs, which have traditionally been overshadowed by large enterprises.

The level of e-records readiness of governments or corporate organisations determines the extent to which they are able to leverage information contained in records for competitive advantage in the digital economy. E-records readiness within governments and corporate organisations can be enhanced if the following measures are taken: where they don't exist, establishing legislations on e-records to formalize their acceptance as official records and legalize their admissibility in courts of law and as evidence of business transactions; creating an enabling environment for e-records management by enacting policies that will guide the systematic management of e-records; and enhancing capacity building by identifying the training needs of records staff and equipping them with skills that enable them to operate in an electronic environment.

REFERENCES

American Records Management Association. (2004). *Information management: A business imperative.* Retrieved on September 19, 2006, from http://www.arma.org/pdf/rim_imperative.pdf

An, X. (2001). *A Chinese view of records continuum methodology and implications for managing electronic records*. Retrieved on October 1, from http://www.caldeson.com/RIMOS/xanuum.html#

Black, H. C. (1995). *Black's law dictionary: Authenticum*. St. Paul: West Group.

Bridges.org. (2001). *Comparison of e-readiness assessment models: Final draft, 2*(13). Retrieved on July 16, 2003, from http://www.bridges.org/eredainess/tools.html

Burness. (2004). *E-Business factsheet-legal freedom of information (Scotland) act 2002*. Retrieved on November 22, 2008, from http://www.scottish-enterprise.com/publications/freedom_of_information_-_dec04.pdf

Case Studies, C. S. R. (2004). *Level playing field for SMEs*. Retrieved on November 22, 2008, from http://www.article13.com/A13_ContentList.asp?strAction=GetPublication&PNID=989

Cashmore, C., & Lyall, R. (1991). *Business information: Systems and strategies*. London: Prentice Hall.

Collins Paperback English Thesaurus. (3rd ed). (1995). *Authenticity.* Scarborough: Harper Collins Publishers.

Collins Pocket English Dictionary. (1996). *Authentic*. Glasgow: Harper Collins Publishers.

Curie, D. (2006). *SME engagement with digital communications services*. Retrieved on November 22, 2008, from http://www.ofcom.org.uk/research/cm/smeengagement

Department for Administrative and Information Services. (2002). *Introduction to electronic records management.* Retrieved on October 23, 2006, from http://www.archives.sa.gov.au/files/management_standard_introelecrecmgmt.pdf

Department of Constitutional Affairs. (2002). Code of practice on the management of records under section 46 of the freedom of information act 2000. [Online]. Retrieved on November 20, 2008, from http://www.foi.gov.uk/reference/imprep/codemanrec.htm

DFID & University of Birmingham. (2003). *Civil service reform*. Retrieved on July 10, 2005, from http://www.grc-exchange.org/g_themes/cc_csr.html

Ford, W., & Baum, M. (1997). *Secure electronic commerce: Building the infrastructure for digital signatures and encryption* (p. 26). New Jersey: Prentice Hall.

Government of South Australia. (2005). *Department of Administrative and Information Services: Across-government records management strategy December 2004 to June 2010*. Retrieved on July 16, 2008, from http://www.archives.sa-gov.au/management/index.html

ICA Committee on Electronic Records. (2005). *Guide for managing electronic records from an archival perspective*. Retrieved on May 11, from http://www.ica.org/biblio/cer/guide_eng.html

InstallShield DevStudio. (2003). *Digital certificates*. Retrieved on January 1, 2007, from http://helpnet.installshield.com/robo/projects/HelpLibDevStudio9/IHelpReleaseDigitalCertificate.htm

Internal Revenue Service. (2000). *VeriSign*. Retrieved on March 20, 2000, from http://www.verisign.com/press/%20rporate/irs_bck.tml

IRMT. (1999). *The management of public sector records: Principles and context*. London: IRMT.

IRMT. (2003a, January 27-31). A summary report on the IRMT / World Bank evidence-based governance in the electronic age. *Global Forum on Electronic Discussion on Information Technology* (p. 8).

IRMT. *(2003b)*. E-Record readiness: Establishing electronic records as a component of electronic government. A proposal to the commonwealth secretariat public sector information program *(pp. 1-6). London: IRMT.*

IRMT. (2004). *The e-records readiness tool*. London: International Records Management Trust.

Iron Mount Digital. (2007). *SMEs believe disaster recovery is just for the big boys*. Retrieved on November 18, 2008, from http://www.ironmountain.co.uk/news/2007/impr05032007.asp

ISO. (2001). *Information and documentation: Records management part 1 general* (pp. 1-20). Geneva: ISO.

Jutla, D., J. Dhaliwal, and P. Bodorik (2002)... Internet Research, 12(2), 139–164. doi:10.1108/10662240210422512

Keen, P., & Balance, C. (1997). *Online profits: A manager's guide to electronic commerce.* Boston: Harvard Business School Press.

Kemoni, H. N. (2008). Theoretical framework and literature review in graduate records management research. AJALIS, 18(2), 17–35.

Lawoye. T. A. (1986). *Records and information management at Nigeria industrial development bank limited (NIDB).* Lagos: University of Ibadan.

Lin, L. S., C. K. Ramaiah, and P. K. Wall (2003). Problems in the preservation of e-records. Library Review Journal, 52(3), 117–125. doi:10.1108/00242530310465924

Lipchack, A., & McDonald, J. (2003). *E-Government and e-records: E-Records readiness and capacity building. Discussion paper.* Retrieved on January 27, 2003, from http://www.irmt.org/download/DOCUME~1/GLOBAL/discussionpaper,pdf

MacNeil, H. (1998). *Trusting records: The evolution of legal, historical, and diplomatic methods of assessing the trustworthiness of records from antiquity to the digital age.* Unpublished doctoral dissertation, University of British Columbia.

McKemmish, S. (1998). *Yesterday, today, and tomorrow: A continuum of responsibility.* Retrieved on January 10, 2004, from http://www. Sims.monash.edu.au
Mcguire, A. (2008). *Public services industry review & SMEs participation.* Retrieved on November 22, 2008, from http://www.whatdotheyknow.com/request/public_services_industry_review

Millar, L. (1997). *Principles of records and archives management.* London: ICA/IRMT.

Ministerstvo, V. (2002). *Analysis of the current state and development trends in the area of information technology and the Internet.* Retrieved on January 1, 2007, from http://www.mvcr.cz/odbor/bezp_pol/english/dokument/anal_eng.html

Mutiti, N. (2001). The challenges of managing electronic records in the ESARBICA region. ESARBICA Journal, 20, 57–61.

Mutula, S. M., & Wamukoya, J. (2007). *Web-based information management: A cross disciplinary approach.* Oxford: Chandos Publishing.

National Archives of Australia. (2006). *Record keeping.* Retrieved on October 1, 2006, from http://www.naa.gov.au/recordkeeping/

Office of the Public Sector Information. (2000). *Freedom of information act 2000 (c. 36): Access to information held by public authorities*. Retrieved on November 18, 2008, from http://www.opsi.gov.uk/acts/acts2000/ukpga_20000036_en_2#pt1-pb1-l1g1

Robek, F., Brown, F., & Stephens, D. (1996). *Information and records management: Document-based information systems.* LA, USA: McGraw Hill.

Sebina, P. M. M. (2006). *Freedom of information and records management-a learning curve for Botswana.* Unpublished doctoral dissertation, University of London, School of Library, Archive and Information Studies, University College London.

Shepherd, E., & Stevenson, A. (2008). *The impact of the UK freedom of information act on records management in the public sector*. Retrieved on November 20, 2008, from http://www.ucl.ac.uk/slais/research/icarus/foi-impact/

Shepherd, E., & Yeo, G. (2003). *Managing records: A handbook of principles and practice.* London: Facet Publishing.

Tough, A., & Moss, M. (2006). *Record keeping in a hybrid environment: Managing the creation, use, preservation, and disposal of unpublished information objects and in context.* Oxford: Chandos Publishing.

United Nations Commission on International Trade Law. (1996). *UNCITRAL model law on electronic commerce with guide to enactment, with additional article 5 bus as adopted in 1998*. Retrieved on March 20, 2000, from http://www.uncitral.org/english/texts/ electcom/ml-ec.htm

University of Pittsburgh. (2004). *Models for action: Developing practical approaches to electronic records management and preservation.* Retrieved on October 13, 2004, from http://www.ctg.albany.edu/publications/journals/models for_action?chapter=3

Upward, F. (1998). *Structuring the records continuum – part one: Postcustodial principles and properties*. Retrieved on October 25, 2006, from http:/www.sims.monash.edu.au/research/rcrg/publications/recordscontinuum/fupp 1.htm

U.S. Department of Justice. FBI. (2008). *FOIA request instructions*. Retrieved on November 18, 2008, from http://foia.fbi.gov/foia_instruc.htm

World Bank. (2006). *Why records management? Records management as a key support for development effectiveness*. Retrieved on October 4, 2006, from http://web.worldbank.org/WBSITE/EXTERNAL/EXTABOUTUS/EXTARCHIVES/0,contentMDK:20033283~pagePK:36726~piPK:437378~theSitePK:29506,00.html

Yusof, Z., and R. Chell (2000). The records lifecycle: An inadequate concept for technology-generated records. Information Development, 16(3), 135–141. doi:10.1177/02666660042404 13

Section 6
Information Needs and Capacity Building in SMEs

This section consists of Chapters 14 and 15 (Information Needs and Access in SMEs and Capacity Building in SMEs, respectively). Section 6 serves two main purposes, namely to demonstrate the difficulties that SMEs face when searching for information to meet their business needs and to try and show how capacity building could empower SMEs that are able to recognise their information needs, identify relevant sources of information, and apply such information.

Chapter 14 highlights why policy developers and managers should find it important to understand the information needs of SMEs. This chapter therefore covers information management in SMEs, information services and access in SMEs, information use by SMEs, factors affecting information seeking in SMEs, challenges of information seeking in SMEs, and policy and managerial roles in meeting the information needs of SMEs. The chapter's aim is to show how SMEs can effectively identify and gain access to critical business information (markets, sources of venture capital, legal advice, use of technology, best practices, policies and regulations, etc).

Chapter 15 maintains that one of the main challenges facing SMEs is skills shortages; consequently, it is urgently necessary to make capacity building a real priority. This chapter covers SMEs' capacity building initiatives, education and training, case studies/best practices in capacity building–Europe, North and South America, Asia-Pacific, The Caribbean, and Africa, information literacy needs of SMEs' employees, and the policy and managerial roles in capacity building for SMEs.

Chapter 14
Information Needs and Access in SMEs

INTRODUCTION

To ensure survival in today's competitive business world, small-sized enterprises require access to accurate and relevant information both at the start-up stage and during their day-to-day operations. SMEs generally tend to be information intensive thus various interventions may be required on the part of management to ensure that the information needs and resources of the enterprises are met and well managed. However, Kirk (1999) doubts whether managers are able to exploit the wealth of information that surrounds them. Given the range of SMEs' needs, it is important to ensure that the information they require to maintain their competitive advantage is both accessible and usable. The managers of SMEs, in turn, also require different aspects of information to plan, organise, staff, administer and control activities in

DOI: 10.4018/978-1-60566-420-0.ch014

ways that best achieve the enterprise's objectives. In the digital economy, successful enterprises are expected to produce high technology goods and services; hence high quality information and effective systems are necessary to achieve such ends

Studies on the information needs of SMEs in parts of Africa showcase the range of information required by SMEs, including information on suppliers, potential customers, standards, acts relating to business ethics and practices, business directories, price lists, etc. SMEs also need knowledge pertaining to management, commercialization and information technology; international trade knowledge; and information about foreign markets (Ntsala, 2000:167-171). In Asia, a study on the IT needs of SMEs in Singapore (Hong Kong) and the Philippines found that SMEs require information relating to credit; the technology market and business opportunities; skills and management training; business exchange/matching; technical assistance in project development; consultancy and advisory services; material sourcing; and research and advocacy. SMEs were also found to need information on (The Asian Pacific Economic Cooperation -APEC, 1995).

Legal Assistance Production-Based Technical Assistance Market Information Policy Reforms

The SMEs' further expressed a desire to be informed about the main trends in technology, particularly the (The Asian Pacific Economic Cooperation -APEC, 1995):

- Rise of e-business and the likely growth in demand for online trading services
- Evolving power of broadband
- Potential of ICT in helping SMEs become high-performance organisations.

As can be seen, SMEs need information that would enhance business growth; government information on licensing; information on tenders; information on taxation and tariffs; information on productivity, sales, and distribution; export information; and information on education and training. Further information requirements include:

- Credit, potential investment and business opportunities
- Market trends and new products
- Viable business projects
- Customers, and quality assurance
- Consumer needs, legislations and judicial decisions.

INFORMATION MANAGEMENT IN SMEs

The creation of knowledge in any organisation and its appropriate use and handling are strategic to the competitiveness of the organisation. Relevant information in the hands of qualified individuals leads to better business decisions. In many SMEs, while specific knowledge generation is a constant activity, the importance of its apt management is sometimes dismissed when it could be crucial to the business' very existence (Bhat and Kumar, 2004). It is therefore important to create awareness of the importance of information management as this would encourage SMEs to establish information management systems and utilize information resources to increase operational efficiency. Quality standards also provide a sound foundation upon which SMEs can generate competitive products including information in the global arena. Such standards also provide a valuable benchmark against which to determine the basis of best practices of top companies around the world (The Asian Pacific Economic Cooperation-APEC, 1995; Soliset al., 2001).

Information resources can best be leveraged in organisations if the information function is integrated with the business strategy. This integration is important as it ensures that managers, as domain experts, are able to use and create information and knowledge which get embedded in the innovations, products, processes or services of the organisations. The need for this integration has long been established in countries in the western world (Broadbent, 1992), and a key factor in aligning business strategy with information has been information technology - a tool essential for business success in the global economy. Managers in SMEs, supported by effective information management and information experts, and like their colleagues in large organisations, are well placed to integrate information and business strategies (Kirk, 1999).

Managers in firms view their success as dependent on their organisation's ability to accommodate and manage change and to respond to changes in their environments. The success of their endeavours rests on information's use, creation, storage and distribution. The quality of information (in terms of its accuracy, validity, timeliness, etc), and its accessibility, availability, presentation, and ease of use, are critical for the organisation's success (Abell, 1994:236). An important stage in the information value chain is information distribution through the sharing of information, whether informally or formally. Information sharing amongst SMEs, especially with respect to financial data, requires a regulation authority that can collect and collate statistics across a wide range of financial institutions to a central repository for aggregation, distribution and sharing (New Zealand Ministry of Economic Development, 2003; Choo, 1995:24-26). ISO/IEC 17799, the international standard that sets out how businesses should conduct the management of their information's security requirements, identifies a number of 'critical success factors' that an organization must achieve in order to be successful in managing information security. These include (British Standards Institute BSI, 2005):

- Having policies in place that reflect business objectives
- Using an approach consistent with organizational culture
- Commitment from management
- A good understanding of the requirements
- Effective policy promulgation
- Suitable training and education
- Feedback to ensure continuous improvement.

The effective management of information also requires adding value to information that is already in use, meaning that there should be room for further development. For example, in organizations, information experts might discuss with managers their media preferences, information user strategies, and the barriers they have encountered in using and applying knowledge. The information experts could then begin tailoring information products and services to enable managers to make better decisions, solve problems, think strategically, scan the environment, and carry out other related functions in their work roles. This kind of process helps information users match the information provided by a system with their actual information requirements (Taylor, 1995).

Mutula and Van Brakel (2006), in a study on the e-readiness of SMEs in Botswana, found that the country's SMEs did not have e-records management guidelines that could facilitate effective access to information contained in records within and outside the organisations. The businesses merely relied on tools provided by various software applications to manage their electronic records. Most enterprises kept records in electronic formats, and staff relied on features provided by Microsoft applications for managing and archiving files. In some cases, enterprises made efforts to transfer records to new applications to ensure that they remained accessible at all times. Both the small and the medium-sized enterprises surveyed used a combination of methods to ensure the authenticity, completeness, accessibility and usability of information. Such methods included validation mechanisms; migration of data; regular backups; offsite storage; and the use of anti-virus programmes. Surprisingly, most SMEs in Botswana had LANs, suggesting that they recognised the importance of sharing information and other resources. However, most of the applications implemented on the LANs were basic and for in-house use, e.g. e-mail communication, databases, Microsoft applications, and product information.

Most of the surveyed SMEs considered the issue of security to be a critical component of information readiness. Findings from the SMEs revealed that they employed a combination of methods to ensure the security of their information and computing resources, such as security systems; the monitoring of transactions; systems operational procedures; antivirus programmes; and regular backups. By and large, most medium-sized enterprises tended to have more sophisticated information security structures in place than smaller-sized enterprises, such as firewalls; regular software updates; offsite storage; authentication; encryption; audit trails; and training programmes. This suggests that medium-sized enterprises were more concerned with security than small-sized enterprises, perhaps because they had invested a lot more money in ICTs and had automated most of their business operations. The findings also revealed that for the most part, little priority had been given to establishing information management units or posts within the organizational structures. This is presumably because at the time, information culture was still a low priority and the size of small-sized enterprises did not warrant such posts

INFORMATION SOURCES AND ACCESS IN SMEs

The sources from which information is sought by SMEs depend largely on the size and sector in which they operate. Generally, these sources are both internal and external. Internal sources may include the Internet; handbooks, trade catalogues, head offices, brochures, internally organised workshops/seminars, etc; while external sources may consist of consultants; external workshops/training seminars, personal visits to relevant offices, etc. SMEs need access to timely information in order to draw from immediate trends and developments in the external environment so that the enterprise can respond to changes triggered by social, economic, technological and legislative forces. Information management within organisations ensures that information is directly delivered to where and when it is needed through clearly defined and understood communication channels (Kirk, 1999).

In the information value chain, important stages of activity include information distribution, information organisation, information storage (an organisation's memory or its sum of knowledge and expertise), and information acquisition. Information distribution activities help with the development of fresh insights and novel solutions through information sharing. These (solutions) can then be captured in the organisation's storage activities and made available for later use. Information organisation and storage activities can help SMEs ensure that the organisation's knowledge is made available to individuals or groups in the organisation and is built on or added to as they expand. Lastly, information acquisition involves the evaluation of current sources, assessment of new sources, and the matching of sources to needs. These activities are all critical to an organization's success (Taylor, 1995).

INFORMATION USE BY SMEs

Access to, and the use of information, are important for SMEs to have the capacity to develop their own information systems. Information can, for example, serve as an essential communication and decision-making tool at all levels of management, and also as an invaluable commodity or product. Because individuals and groups have different objectives within unique organisational settings, all attributes cannot be of equal value to each user. Instead, separate but related information systems are needed to deliver the desired attributes to the different users.

Information and its successor, knowledge, have the potential to transform organisations and make them more competitive than could previously be accomplished through physical resources such as land, capital and labour In the words of Drucker (1993:43): "Knowledge is the only meaningful resource today. The traditional "factors of production" - land (that is, natural resources), labour and capital - have not disappeared, but they have become secondary. They can be obtained easily, provided there is knowledge". The importance of information in gaining competitive advantage should be emphasized. A survey of high performing companies in the UK (Owens et al., 1997) revealed that all of them were of the view that information is a valuable asset. A growing number of successful companies are concerning themselves with information management while striving to create an information culture in their organizations to ensure continued success. The potential of information technology and information to transform organisations is especially evidenced in companies that have redesigned their business processes and information systems (Johannessen and Olaisen, 1993). These information systems support decision-making, provide an effective interface between users and computer technology, and provide information for the management of the day-to-day, routine operations of the enterprise and in support of the unique needs of professionals. Dawes (1996), based on a study on inter-organisational information sharing amongst government managers, concluded that information is valuable for better, more integrated planning; policy development and program implementation across agencies; providing more comprehensive and accurate data for decision-making and problem solving; enhancing the productive use of scarce staff resources; and facilitating better interagency and professional relationships.

FACTORS AFFECTING INFORMATION SEEKING IN SMEs

Research into the value and use of information to enhance productivity and informed decision-making in business environments and other organisations started a long time ago, and has since been gaining momentum in step with the evolvement of ICTs. Current practices and ways of retrieving information using technology have been largely built on the earlier works of information professionals.

The ability of an organisation to realize its goals depends on how well the organisation acquires, interprets, synthesises, evaluates, and understands information, and how well its information channels support organisational processes. One of the main factors affecting the use of information has to do with access, which significantly influences both the level of use of an information system and a user's evaluation of the sourced information. The extent to which individuals obtain information indirectly, for example through search analysts who retrieve specific information as per request; or directly as end users, helps in the assessment of their level of competence. Users who obtain information indirectly through search analysts are distanced from the context of the information sources and the retrieval process. Such users are likely to be less familiar with the specific content and commands of various databases. They would therefore not have the opportunity to develop or expand on the familiarity required to retrieve information successfully, inexpensively and conveniently. Because they are removed from the selection and interpretation of online information, indirect users are less likely to believe in causal relationships between the source of information and any improvements in performance. Furthermore, it would be harder for the search analyst to understand the user's task context and know what constitutes effective results to ensure that the information the search analyst provides contributes highly to the user's work performance (Hart and Rice, 1991:461-471).

Generally though, the greater use of information from external online databases appears to be associated with perceived improved work effectiveness. Direct users of information systems have tasks that require them to do at least one of the following: access databases on a regular basis; download and manipulate information; and perform a series of contiguous steps, each of which requires the analytic skills of the user to determine how to execute subsequent steps. Moreover, if the new sources or systems are not highly accessible, conventional alternatives will be used even when their response time is slower and the quality of information is lower. Rice (1987:67) observed that how one accesses the information system, what one's tasks are, what kinds of databases one uses, and the problem solving needs of one's organisation, represent some of the central influences on a person's ability to evaluate and use information.

CHALLENGES OF INFORMATION SEEKING IN SMEs

One of the most notable obstacles limiting SMEs capacity is access to timely, current, relevant and adequate information for informed decision-making. SMEs generally struggle to gain access to important information in a timely manner for improved productivity, profitability, and customer satisfaction. SMEs are confronted with this struggle because they either do not understand what information is relevant or required, and/or do not know how to obtain it efficiently.

A study undertaken by Duncombe and Heeks (2001:32) in Botswana on the information seeking behaviour of SMEs found that they (SMEs) were generally under-performing due to lack of access to relevant information on finance and credit. All over the world, small-sized enterprises are often not well served with the information they need to keep up with local, national and international business trends. In part, this lack of access to information can be attributed to the fact that normally, small-sized enterprises do not have libraries and documentation centres of their own. Furthermore, the information services that are provided to small-sized enterprises tend to not only be scarce, but also fragmented and uncoordinated. SMEs are also handicapped in finding and accessing information because of the low levels of education and business skills of their staff; limited research on the SME sector; poor regulatory frameworks; lack of a comprehensive entrepreneurial strategy; and the poor visibility of small-sized enterprises. Other barriers include:

- Lack of access to technology
- Vulnerability due to cash flow disruptions
- Unfavourable tax regimes, such as value-added tax and skills levies
- Inability of some governments to communicate what incentives are available for emerging entrepreneurs and where to go for assistance.

POLICY AND MANAGERIAL ROLES IN THE INFORMATION NEEDS OF SMEs

SMEs generally have less sophisticated systems than larger enterprises that could enable them to gain access to the information they need. In order to improve access to information and encourage SMEs to extend their market reach, governments should set up business portal links on their websites offering access to information on business-related operations such as tenders, sources of credit, markets, products, services, pricing, tariffs, and legislations. Governments should also strive to establish business information centres that would employ state of the art technology in order to provide a one-stop shop and a full range of management and technical assistance to SMEs. Furthermore, consideration should be given to the establishment of networking initiatives which would increase the contact between SMEs, prospective clients and/or suppliers throughout their jurisdiction and beyond.

CONCLUSION

Most of the information sought by SMEs relate to legal assistance, market information (e.g. trends and prices), policy changes and legislation, technical assistance, access to credit, etc. In order to satisfy these needs, SMEs can make use of different sources of information, such as the Internet, brochures, colleagues in other enterprises, and through workshops and seminars. The information acquired can be put to use in areas such as planning and policy formulation, understanding market trends, undertaking market intelligence, and much more.

As pointed out in this chapter, SMEs face various information seeking challenges, including lack of access to information, the inability to interpret and effectively use information when it is received, and poor information security. Governments should intervene and help SMEs by, for example, making efforts to formulate and coordinate a national programme of policy research and the collection and dissemination of information concerning small-sized enterprises; and encouraging them to regularly carry out information security audits and needs assessments in order to deploy security measures that are reliable and effective. SMEs, in turn, should make attempts to apply the ISO/IEC 17799 international standard on managing information security requirements.

REFERENCES

Abell, A. (1994). Information use and business success: A review of recent research on effective information delivery. In M. Feeney & M. Grieves (Eds.), *The value and impact of information* (pp. 229-253). London: Bowker Saur.

APEC. (1995). *APEC survey of technology information needs and training opportunities for SMEs, APEC Committee on SME.* Manila, Republic of the Philippines: Coordinating Agency, Dept. of Trade and Industry. Retrieved on August 21, 2004, from http://www.actetsme.org/index.html

Bhat, I. S. A., and V. Kumar (2004). A structured approach to knowledge management in SMEs: Towards a successful manufacturing strategy. International Journal of Business Performance Management, 6(3/4), 233–244. doi:10.1504/IJBPM.2004.005630

British Standards Institute (BSI). (2005). *ISO/IEC 17799*. Retrieved on March 4, 2005, from http://www.articsoft.com/iso17799.htm

Broadbent, M. (1992). The emerging phenomenon of knowledge management. The Australian Library Journal, 46, 6–23.

Choo, C. W. (1995). *Information management for the intelligent organisation: The art of scanning the environment*. Medford, NJ: Information Today.

Dawes, R. M. (1996). *Theoretical model of individual agency-to-agency information sharing*. Retrieved on May 6, 2004, from http://64.233.161.104/search?q=Consulting and Audit Canadahe:fqFa7g-WNOAJ:etd02.lnx390.lsu.edu/docs/available/etd-0619103-214616/unrestricted/03CHAPTER2.pdf+Dawes+1996+and+innovation +diffusion&hl=en Drucker, P. (1993). *Post-capitalist society*. New York: Harper-Collins.

Duncombe, R., & Heeks, R. (2001). *ICTs and small enterprise in Africa: Lessons from Botswana* (pp. 1-167). Manchester: University of Manchester, IDPM.

Hart, P. J., and R. E. Rice (1991). Using information from external databases: Contextual relationships of use, access method, task, databases type, organisational differences, and outcomes. Processing and Management, 27(5), 461–479. doi:10.1016/0306-4573(91)90063-R

Johannessen, J. A., and J. Olaisen (1993). The information intensive organisation: A study of governance, control, and communication in a Norwegian Shipyard. International Journal of Information Management, 13, 341–354. doi:10.1016/0268-4012(93)90004-N

Kirk, J. (1999). Information in organizations: Directions for information management. *Information Research, 4*(3). Retrieved on March 3, 2005, from http://informationr.net/ir/4-3/paper57.html

Mutula, S. M., and P. Van Brakel (2006). An evaluation of e-readiness assessment tools with respect to information access: Towards an integrated information rich tool. International Journal of Information Management, 26(3), 212–223. doi:10.1016/j.ijinfomgt.2006.02.004

New Zealand Ministry of Economic Development. (2003). *Bank lending practices to small and medium sized enterprises*. Retrieved on March 5, 2005, from http://www.med.govt.nz/irdev/ind_dev/access-to-finance/lending-practices/lending-practices-06.html

Ntsala, M. (2000, October 26-27). Information delivery systems for small business. In *Proceedings of First Biannual DISSAnet Conference* (pp. 167-176). Pretoria: The Centre for Information Development, University of Pretoria.

Owens, I., T. D. Wilson, and A. Abell (1997). Information and business performance: A study of information systems and services in high performing companies. Journal of Librarianship and Information Science, 29, 19–28. doi:10.1177/096100069702900103

Rice, R. E. (1987). Computer mediated communication systems and organizational innovation. The Journal of Communication, 37(4), 65–94. doi:10.1111/j.1460-2466.1987.tb01009.x

Solis, E. L., S. S. Rao, and S. Ragu-Nathan (2001). The best quality management practices in small and medium enterprises: An international study. International Journal of Manufacturing Technology and Management, 3(4/5), 416–443. doi:10.1504/IJMTM.2001.001421

Taylor, R. S. (1995). *Value-added processes in information systems*. Norwood NJ: Ablex.

Chapter 15
Capacity Building in SMEs

INTRODUCTION

The preceding chapter demonstrated that one of the more pressing challenges facing SMEs is the lack of adequate skills, which makes capacity building a critical preoccupation. The 2003 WSIS highlighted the need for capacity building to achieve an information society where there is rapid growth and the wide spread use of information, and where people have the necessary literacy competencies to appreciate what information is needed, where to get it, and most importantly, how to use it. Appropriate levels of education and Internet familiarity are necessary for digital commerce to be viable. Digitally enabled consumers and businesses are the basic cornerstones of the digital economy. Without the necessary ICT skills, firms would find it difficult to penetrate global markets. Notably, labour-intensive businesses that

DOI: 10.4018/978-1-60566-420-0.ch015

support digital commerce, such as call centres or KM outsourcing providers, move quickly into markets where skills are available to support their operations.

SMEs generally lack the human technological resources needed for e-commerce because they focus on day-to-day operations and lack the time to understand the benefits of new technologies. When they are aware of the potential benefits of e-commerce, they find that they still need to invest in the expertise of qualified personnel. Firms that adopt the Internet and e-commerce are likely to have within them persons who have a reasonable amount of knowledge on specific technologies used within the enterprise and/or technology in general. Internal technological capabilities matter because the adoption of e-commerce cannot succeed without a solid understanding of the e-commerce business model. Furthermore, dependence on external ICT support services for system maintenance could be dangerous because firms that conduct e-commerce potentially risk contracting computer viruses or suffering other system failures at virtually any point in time, leading to losses or other problems in the business if help isn't immediately on hand. Some of these matters could be quickly and effectively resolved if internal technicians or ICT support services were available. Hence, the existence of local ICT skills is an important factor, especially for small offices in remote areas or emerging economies where ICT services are generally scarce and costly.

SME CAPACITY BUILDING INITIATIVES

Porter (1990), in a study that focused on Germany, Japan and Korea, established that education and training are decisive in determining national competitive advantage. Nations that invest heavily in education have advantages in many industries that can be traced back to human resources. Highly competitive industries have often made significant investment in the education and training of their personnel. Infact, education and training constitute the single greatest long–term investment in any organization and at all levels of government. Without knowledge, even nations that are well endowed with natural resources cannot compete effectively on the global arena.

Capacity building in SMEs is increasingly being perceived by governments the world over as a sustainable empowerment tool in the global economy, although it would seem as though most small-sized enterprises do not put much premium on training. For example, in a UK study on SMEs, Gray and Lawless (2000) found that the small-sized enterprises were, on average, generally averse to adopting a formal staff development policy. They attributed this to a number of external factors, such as the scale effects on resources and time, relative organisational simplicity, and the need to remain flexible (hence the dominance of a 'react to need' approach among the SMEs). The authors observed that most small-sized enterprises viewed the formal development of a capacity building policy to be of low priority, whereas most medium-sized enterprises felt that staff capacity building should be high on their agenda. Furthermore, the level of formality, planning, and even successful nature of staff development appeared to be linked to the complexity of the business. Firms with written, explicit staff development policies - mainly medium enterprises or large firms - were more likely to benefit from most forms of training.

It has already been mentioned that in order to fully exploit the benefits of ICT, the adequate development of human capacity is necessary. Furthermore, creating awareness of the abilities of ICT to improve the lives of people by circumventing traditional obstacles, such as distance and time, is necessary within SMEs. McConnell International (2000) opines that the world faces an incredible shortage of ICT skills, particularly in four areas:

- Managers capable of completing complex technological projects
- Policy analysts who understand government
- Local content creators
- Software, hardware and communication engineers.

SMEs often tend to have a strong demand for technical personnel to meet the needs for service and support, as well as for network deployment, management and maintenance. It is therefore important to effectively procure human, technical and financial resources if [a country] is to meet its technological needs. An important approach to human capacity building is the tailoring of the immigration policy to welcome skilled high-tech workers from abroad. Furthermore, identifying the requirements of the organization, implementing appropriate training programs, and educating people on the procedures and mechanisms that would lead to the best results in terms of the technologies, can have a significant impact on the adoption and use of ICT in SMEs. Developing human capital also necessitates incentives to attract, train and retain the best technical and managerial talent.

EDUCATION AND TRAINING

Enhancing education and training through the effective application of ICT, both as a classroom tool and a subject in its own right, should become one of the main priorities of governments the world over. It is important for ICT related skills to be developed through the traditional education system because SMEs tend to provide little or no formal training themselves, and expect to either hire qualified staff, or to have the staff train themselves. Education is fundamental because it supports the national development or ICT agenda of any country. A sustainable ICT programme that contributes significantly to a country's development can only be achieved if there is an existing suitably trained workforce and an education system that continually supplies skilled, innovative and entrepreneurial professionals. As far as the national investment in education is concerned, business enterprises should have a significant role to play in contributing to citizens' knowledge and skills in addition to that of their own personnel. Companies that invest more heavily to attract, train and retain employees can help raise the overall level of workforce productivity. In many instances, they also provide people with the skills they need to participate more fully in the information economy.

Training programmes developed for SMEs should meet some level of standardisation. For example, if SMEs wish to communicate locally, nationally and internationally, they need to adopt appropriate standards that would allow interconnectivity and compatibility with global networks. Standards also provide a benchmark for measuring performance and enhancing the effectiveness of service offerings and associated support. As new technologies emerge and policies and procedures are developed to accommodate new associated services, standards become effective mechanisms to ensure that appropriately trained staff can support them. Operational standards, service level agreements and other best practices are good indicators of how customer services can be improved and an organization's performance raised to reach internationally accepted levels. The European Office of Crafts, Trades and SMEs for Standardisation (1994) found that often, SMEs considered standards to be a burden specifically made by large groups for large groups; and thought that they (standards) did not apply to them. By applying standards, SMEs can ensure that their products and services are consistent, compatible, safe and effective.

The successful adoption of ICTs requires an educated, technically comfortable populace, whose technical literacy is predicated on their ability to read and write well. Developing human capital through the implementation of ICT in primary, secondary, and post secondary education and training is the place to start if SMEs are to gain the critical mass of skills needed for the digital economy. The level of digital awareness and requisite skills are also crucial to the deployment and use of various ICTs. **For this reason,**ILO (2001) states that countries with the right mix of skills, infrastructure and policies stand a better chance of actively and successfully participating in global markets. Therefore, in order for maximum gains to emerge, the development of essential ICT skills is necessary, because without such skills, the technologies cannot be maintained or adapted to local use. Literacy and education cannot be leapfrogged, and both are vital to reaping the greatest advantages from the emerging digital economy.

The promotion of education and literacy in general and digital literacy in particular is a challenge facing all countries, and the educational differences within and between countries is responsible for the different rates of penetration of ICT and Internet usage. Access alone is not sufficient; workers need to be trained in the substance of new technologies and their respective uses. Furthermore, the adoption of ICT in SMEs creates two types of skills needs. The first is related to essential basic skills, such as the ability to learn to communicate and to analyse and solve problems. Beyond such skills are technical skills, which are related to ICT itself, the needs of which extend beyond the ICT sector to the economy as a whole. Wherever technologies are most widely used, skills shortages, particularly in the areas of technical support for both hardware and software applications are acute. This is a drawback to the economic growth of enterprises that would otherwise be adept in the technologies' applications (ILO, 2001).

The widespread use of emerging technologies, both at work and at home, demands flexible workers who are able to keep up with technology, are self-directed and knowledgeable about the world, and who can facilitate and perform cross-cultural communication and transactions. Moreover, the education of this new workforce requires that emphasis be placed on information access, problem solving, analysis, evaluation and decision making. Among developed and developing countries, there is likely to be a wide variation in the level of awareness of the application of ICT within the SME sector (Zakaria, 2001). In Canada, most firms are aware of the considerable benefits of e-commerce. However, in South Africa very little was known about the technology needs of the SME sector (Agriculture and Food Canada, 2001; Gordon, 2003). At the time, the use of the Internet for commerce was still in its infancy, and there was therefore a considerable lack of ICT-related knowledge. Ramsey et al. (2003:252), in an Irish cross border study on e-opportunities in the SME service sector, concurred that one of the main problems facing SMEs was lack of awareness, and found that quite often, SMEs equated having a website to electronic commerce. Such SMEs would setup a website and then not know what to do with it. In essence, the website sits there as a reminder of their failure.

CASE STUDIES OF CAPACITY BUILDING

Various capacity building initiatives are being undertaken by governments and the private sector the world over to enhance ICT awareness, literacy, and skills development.

a) Europe

The eEurope Action Plan is a European Union strategy aimed at accelerating, reinforcing and modernising member countries' economies by encouraging capacity and institutional building through investment in people and skills. The EU member states are taking advantage of the growing use of ICTs and the spread of the Internet to put very powerful tools within the reach of citizens, governments and both large and small-sized enterprises sectors. This development has resulted in profound changes in the initial organisation of governments and businesses, and also in skills requirements. The European Union recognises that the shortage of ICT and business skills in Europe have prevented SMEs from fully realizing their employment potential (eEurope+, 2003:16). The EU member countries also recognise that productivity gains depend largely developing higher levels of skills and taking advantage of the greater flexibility made possible by new technology. The member states are therefore striving to increase the number of trained ICT and e-business experts in their midst in order to enhance the competitiveness of their economies. Across the workforce, it was also found that broader action is necessary to promote digital literacy through lifelong learning and create awareness of the opportunities of ICT training. Member countries are presently implementing a common European diploma that teaches basic digital skills to stimulate the uptake of ICTs in the economies.

In 2000, the European Union estimated that there was a shortfall of 700,000 IT specialists, and this number was expected to rise to 1.3 million by 2003 in the region (Mizzi, 2004:15). Consequently, European Union members were lobbying for the use of structural funds in enabling widespread Internet access and the training of teachers in Internet literacy. The EU is further re-skilling its workforce in order to improve its operation in the information society. The e-Europe Action Plan envisaged that by the year 2003, all adults would have been provided with the digital skills they need to improve their employability and overall quality of life. In addition, all member countries were expected to ensure that all their universities offered online access to students and researchers in order to maximise the quality and efficiency of learning processes and activities (The Commission of the European Communities, 2002:14).

Individual countries in the EU have been making efforts to implement capacity building initiatives. Malta's government has, for example, put in place many initiatives to enhance ICT literacy. The region's government has been collaborating with the private sector to accelerate and upgrade those initiatives aimed at providing widespread digital literacy and producing much-needed IT specialists to fill the sector's labour shortage. In 1995, one such collaborative initiative saw the education authorities embark upon a widespread programme for the dissemination of PC's to schools, starting from primary grades, to ensure that the necessary IT skills would be widely available in the country (Mizzi, 2004:15). In the UK, a study on the relationship between the diffusion of ICTs and changes in skills within business organisations, found that the changes were also correlated with productivity (Hwang 2004), Education and training were found to be important factors in modifying skills to cater for the rapid expansion of ICTs. The British government has made a commitment to ensuring that the country is a leader in the global information age, and one way it aims to achieve this is by encouraging the widespread development of information and computer skills in the public and private sectors and in the population as a whole (Horrocks and Haines, 2004). In Turkey, another EU member state, the government has been making efforts to create awareness of the benefits of access to the right information at the right time via the Internet, especially with regard to finding information relating to financing SMEs (Gungen, 2003).

b) North and South America

The North American region is making great strides in its initiatives to enhance capacity building for the benefit of SMEs. In the United States, Lee et al. (1994:5) found that one of the greatest barriers facing the SMEs' export business was the sector's lack of awareness about where to get information on how to begin the export process. The authors observed that without strong education and encouragement, a typical United States SME tends to avoid the unknown of doing business in foreign markets. Another study in the United States on the e-readiness of SMEs in the agricultural sector found that there is a growing acceptance of the Internet by the SMEs as a medium to provide information or locate suppliers, and also as a way to transact business electronically and thereby reduce transaction costs. The growing use of the Internet for business transactions is attributed to an increased awareness of the benefits of ICTs. In 2001, although e-commerce was an important component of e-readiness in Canada, a very small fraction (6%) of the country's SMEs was using the Internet for e-commerce. One reason advanced for this at the time was the low level of knowledge about the Internet, which prevented most firms from realising the benefits of e-commerce. Complex technology and the lack of skills are also identified as barriers to the adoption of e-commerce (Agriculture and Food Canada, 2001).

The Canadian government recognises the important role of education and training in developing digital literacy to reap the benefits of ICTs. Emphasis has consequently been placed on awareness initiatives relating to ICT and knowledge and how they can be utilised to access information and services (Agriculture and Food Canada,

2001). The impartation of appropriate education has also been emphasised to help the population adapt and reshape ICT to create new content or seek specialised employment within the ICT sectors. In 1996, the government was considering how entrepreneurship and business skills could be integrated into schooling. In that year, the government, in partnership with the private sector, launched the Student Connection Programme, which hired and trained university and college students as student business advisors to SMEs across Canada. The programme provided practical, onsite IT training tailored to suit the needs of everyone, from the beginner to the advanced user. Through the programme, companies were expected to improve their personnel's awareness of the value of incorporating electronic commerce in their day-to-day business activities (Consulting and Audit Canada, 2004:31-32). ICT skills also lead to investment benefits. In Costa Rica, for example, the educational level of the workforce was considered to be a vital factor in Intel's decision to relocate a semiconductor production facility to the country.

c) Asia-Pacific

A study on SME trade and investment in countries belonging to the Asia-Pacific Economic Cooperation (APEC), i.e. Singapore, Thailand, the Philippines, Hong Kong, Indonesia and Malaysia, noted that alongside the principle obstacles facing SMEs wishing to participate in global markets were: poor knowledge and experience with regard to the mechanics of the export industry; lack of personnel; limited overseas market intelligence; and insufficient management expertise, experience and training (Lee et al., 1994:6). The Economist Intelligence Unit's annual global survey of e-readiness for both 2003 and 2004 revealed that although India and China were some of the world's biggest economies with a world-class cadre of software programmers, booming outsourcing industries, and the world's fastest growing Internet populations; though they performed poorly in e-readiness rankings (Economist Intelligence Unit and IBM Corporation, 2004). India's leadership in software programming and outsourcing may be attributed in part to the fact that the government recognises that there are many factors which together are most critical for the development of the country among them its citizens. The citizen component consists of the skills sets of a nation's populace, such as their literacy levels and attitude towards the adoption of ICT (The Department of IT eTechnology Group-India, 2003).

d) The Caribbean

Within the Caribbean region, there is a growing use of ICT in the SME sector, and governments have shown commitment by investing in education to meet the workforce demands of the digital economy. For example, in Jamaica, the increased use of ICT has been attributed to the growing awareness of its uses (Southwood, 2004). In 2000, Trinidad and Tobago's government expenditure was estimated at US$ 250 –US$ 325 per capita, placing it in the top tier of 2004's world ranking of public expenditure on education. High public expenditure on education demonstrated the country's commitment to elevating and maintaining the knowledge of its citizens (Southwood, 2004).

e) Africa

An e-readiness ranking of African countries with regard to human capital, categorised on a continuum of five levels, i.e. medium-high levels, medium levels, low-medium levels and low levels, found that on a gross enrolment ratio, only South Africa was ranked in the medium levels, while the rest of the countries were ranked in the low levels category. In terms of the enrolment of IT students in tertiary education, countries in the high levels included: Botswana, Cameroon, Mauritania, Nigeria and Tanzania. Countries in the medium high levels were: Angola, Burundi, Ethiopia, South Africa, Zimbabwe, Ghana, Gabon, Eritrea, Senegal, Guinea, Cote d'Voire, Madagascar, Niger, and Togo. The medium levels consisted of countries such as: Benin, Kenya, Swaziland, Zambia, Uganda, Lesotho, Mali, Chad, and Burkina Faso. Mauritius was ranked in the low medium levels, while Guinea Bissau was in the low levels category. These results implied that by and large, most countries in Africa were not e-ready in terms of their human capital (Docktor, 2004:14-33).

A study carried out in 2003 by Maksoud and Youseff (2003) on the use of ICTs in SMEs in Egypt found that unqualified personnel, followed by lack of awareness, were the topmost barriers to the use of ICTs in these firms. The market in Egypt was reported to be at the early stages of consumer adoption, and the suggestion was made that by catalysing awareness, the rate of adoptions could be accelerated. At the time, the government had included computer learning in most schools and universities in order to create a skilled base of entrepreneurs. The authors noted that the country had made good progress in fostering IT skills. They recommended that ICT awareness should begin in schools and continue through to university, and trained professionals should be hired to provide instruction. Such professionals should be specialised in ICT tools for businesses so that they can assist in raising the entrepreneurial culture of the country's students (Maksoud and Youseff, 2003).

The SADC e-Readiness Task Force (2002:2), in a review of the strategy of SADC member countries, pointed out that a paradigm shift was necessary in the social attitudes and approach to skills levels and technology within most of the urban and rural communities in the region, and within both private and public sector organisations. The review noted that member states needed to implement innovative projects to increase the awareness and use of ICTs in everyday life. The Task Force emphasised the role played by primary, secondary and tertiary education as e-readiness enablers, as well as the need for ICT-based skills and the incorporation of technology-based instruction in the region's curricula and teaching methods. The Task Force also recommended for tertiary education to be further developed at both local and regional levels in order to foster and grow the ICT skills base within the region. The review found that national educational policies need to emphasise ICT literacy, including curricula reforms. Likewise, teaching policies need to emphasise the additional training of teachers as a primary requirement. It was also recognized that there is a need to make education and skills capacity building a national priority, with attention from the top agencies of government.

The SADC e-Readiness study also reported that entire communities within the region need to be exposed to the benefits of ICT to instil positive attitudes towards its adoption. This could be achieved through different communication initiatives, such as a national awareness campaign. The study ranked Botswana first in terms of the ratio of school learners and tertiary students per population (%); followed by Namibia, South Africa, Zimbabwe, Seychelles, Swaziland, Malawi, Zambia, Lesotho, Mauritius, Tanzania, DRC, Mozambique and Angola. It was found that education at tertiary level was considerably low in the entire SADC region. By implication, this would mean that there wasn't a strong education system for the development of the skills necessary for participation in the digital economy. In a study by Consulting and Audit Canada, which aimed to promulgate a national ICT policy for Botswana, it was found that among the major problems facing ICT companies in the country was the problem of limited human resources (Maitlamo, 2004:6). The study noted that while access to adequate and top-level human resources is crucial to ICT sector development, there were no PhD level employees in any of the companies surveyed. The report noted that PhD level entrepreneurs combine in-depth knowledge with research and business skills - a combination that has proven to be the cornerstone for innovative thinking and new developments that have taken place in this sector around the world.

On the whole, the developing world, including most countries in Africa, is constrained by the lack of ICT skills necessary to effectively take advantage of the digital economy. This was underscored by SIDA (2003) when it pointed out that although ICT was gaining ground in the developing world, there was still a critical lack of technical personnel and computer literacy.

INFORMATION LITERACY NEEDS OF SME EMPLOYEES

There are a number of information literacy models that attempt to explicitly define the types of competencies an individual should have in order to effectively participate in the digital economy. For example, the American Library Association (ALA) outlines information literacy (IL) competencies that could be used in the design and implementation of education and training programmes. These standards define IL as a set of abilities that require individuals to: determine which information is needed; access the information effectively and efficiently; evaluate information and its sources critically; incorporate selected information into one's knowledge base; use information effectively to accomplish a specific purpose; understand the economic, legal, and social issues surrounding the use of information; access and use information ethically and legally; and manage and maintain information (American Library Association, 2003). Other equally important competencies relate to knowing how to use different information resources; organizing information; providing access to information; searching for information; and gathering, evaluating and using information.

IT skills are also needed to participate in the digital economy, and these skills include the ability to use computers, software applications, databases and other technologies to achieve a wide range of work-related goals. SCONUL (Standing Committee of National University Librarians, UK) is an example of an information literacy model that defines key competencies for individuals to live and operate by for effective participation in the electronic age. SCONUL defines information literacy as "an understanding and set of abilities enabling individuals to recognize when information is needed and have the capacity to locate, evaluate and effectively use the needed information" (Horrocks and Haines, 2004). SCONUL's model is built on seven pillars, namely the ability of an individual to recognize a need for information; distinguish ways in which the information 'gap' may be addressed; construct strategies for locating information; locate and access information; compare and evaluate information obtained from different sources; organize, apply and communicate information to others in ways appropriate to the situation; and synthesize and build upon existing information, contributing to the creation of new knowledge.

Information literacy is an essential life skill for the employees of any organization because it enables them to be more effective in their jobs and as citizens. Information literacy competencies should therefore cover a range of IT and information retrieval and management-related skills, such as sifting through basic and advanced bibliographic and full-text databases, Internet browsing, and using the full range of Microsoft Office applications. The World Summit on Information Society (WSIS, 2003) underscored some aspects of information literacy and the skills needed for one to operate in the networked world, and the most prevalent of these is the ability to create, receive, share and utilize information and knowledge in any media and regardless of frontiers.

Mutula and Van Brakel (2006) integrated e-readiness model covers aspects of information literacy categorised under five dimensions namely: Human Resources Readiness, ICT Readiness, Information Readiness, Enterprise Readiness, and External Environment Readiness. The Human Resources Readiness component of the model outlines the information literacy and skills competences believed to be necessary for participation in the networked world, including being able to determine one's own information needs; understanding the economic and legal issues of information; evaluating information and its sources; organising information; managing and maintaining information; accessing, analysing, and using information; and understanding the benefits of information. Other aspects of this component are abilities demonstrated through:

- Success rates in searching and retrieving information
- Levels of awareness of ICTs
- ICT-related educational qualifications and skills
- Positive attitudes towards the use of ICTs
- Level of use of ICTs.

With regard to Information Readiness, the model focuses on the ability of enterprises and individuals alike to define their information needs; provide access to various sources of information; access information through electronic media; provide mechanisms to collect, store and retrieve information; avail standard formats for information organization, storage and retrieval; offer adequate information retrieval tools; facilitate the free and rapid flow of information within the enterprise; and provide a means of sharing information within and outside the enterprise. Other aspects of the Information Readiness component include: ensuring the authenticity and completeness of information; compliance with information management practices (e.g. backup); provision of information management functions; and information management functions' integration in business processes (Mutula and Van Brakel, 2006).

The ICT Readiness component of the model proposes the following indicators: pervasiveness of ICTs in the enterprise; Internet connectivity; access to the Internet and the World Wide Web by staff; high bandwidth availability for network access; quality of connectivity to the network; website availability for business promotion; mechanisms for keeping informed of developments in ICTs; and the level of online security. Lastly, in terms of the External Environment Readiness component, the model proposes enabling legislative and regulatory frameworks to address intellectual property rights for e-commerce; improving the quality of telecom services, including adequate bandwidth; increasing ICTs affordability with regard to service providers; facilitating freedom of information; improving the quality of nationwide ICT infrastructure; ensuring the adequacy and reliability of the national power grid; maintaining the security and reliability of the network to support e-commerce; and developing an enabling e-business environment. The External Environment Readiness component also includes: the level of government support; credit facilities available to enterprises; universal access; national information policy; affordability of bandwidth; providing an enabling taxation regime; and the availability of a competition and consumer protection policy.

POLICY AND MANAGERIAL ROLES IN CAPACITY BUILDING FOR SMES

Governments have profound roles to play in promoting and facilitating capacity building within their respective jurisdictions. Efforts are also needed on the part of the managers of SMEs to enhance skills development. The model of the EU, i.e. increasing the number of trained ICT and e-business experts in the member states to enhance the competitiveness of their economies, is worth emulating by other countries. Broader action is also necessary to offer lifelong learning in ICT-related skills across the workforce and to focus attention on the opportunities of ICT training. Considerations should be given to implementing basic digital skills' programmes to stimulate the uptake of certified training and its recognition, promote a culture of learning, and create training centres that cater for demand driven information. Governments should put in place mechanisms for the effective procurement of human, technical and financial resources and enhance their research and development capacity in order to meet the technological needs of their nations. As a matter of policy when awarding tenders to external bidders, and especially in the developing world, governments should include a requirement for international companies to work with local companies in order to achieve skills and technological transferrals and thus obviate the situation where the country will remain dependent on foreign experts.

Governments should develop a prudent approach to capacity building, for example by tailoring the immigration policy to welcome skilled, high-tech workers from abroad. Likewise, identifying the organization's requirements, implementing appropriate training programs, and educating people on the procedures and mechanisms for better IT utilization can have a significant impact on the adoption and use of ICT within SMEs. It is especially important for ICT related skills to be developed through the traditional education system because SMEs have a tendency to provide little or no formal training, and expect either to hire qualified staff or for staff to train themselves. Governments should also put in place institutional and human resource strategies that address national skills shortages and promote competition in order to stimulate new investment.

Finally, SME managers should put in place mechanisms to enhance the skills' capacities of their staff, especially in digital competencies such as web design, knowledge management, and business intelligence. In particular, firms should put in place explicit staff development policies in order to harness the potential benefits of ICTs. Clearly, ICT capacity building should focus on training managers who are capable of completing complex technological projects; policy analysts who understand government regulatory frameworks; local content creators; and software, hardware and communication engineers.

CONCLUSION

Internationally, capacity building is increasingly being seen as a vital component in initiatives that aim to help SMEs actively participate in the global economy. A lot of emphasis is being placed on education and training, especially on including ICT as a classroom tool and as a subject. Within the European Union, member states are encouraging capacity building through their investment in people skills. In North America, the US and Canada have put more emphasis on creating awareness of the benefits of ICTs in business. Canada established a Student Connection Programme that focuses on training university students as business advisors for SMEs. In the Caribbean, the government of Jamaica has invested heavily in education to meet the digital labour needs of this century. Within the South African Development Community (SADC), member states are at different stages of installing ICT infrastructure and developing capacity among other efforts to benefit from the digital economy.

Capacity building that integrates information literacy competencies is central to any meaningful education and training initiatives, especially with regard to the use of ICT. A number of models have been discussed in this chapter that provide benchmarks which can be used to develop an effective information literacy education programme, especially with regard to digital literacy in the global economy. Such models include but are not limited to the American Library Association (ALA) model, the British standard by SCONUL, the WSIS Declaration of Principles (defines the kinds of skills needed in the digital age), IFLA's model (provides standards for information literacy), and Mutula and van Brakel's (2006) integrated e-readiness model.

REFERENCES

Agriculture and Food Canada. (2001). *The e-readiness of Canadian Agriculture and Food Canada SMEs*. Retrieved on March 21, 2004, from http://www.agr.gc.ca/misb/fb/food/ecommerce/eready/introduction_e.html

American Library Association. (2003). Information literacy competency standards in higher education. Retrieved on November 7, 2004, from http://www.ala.org/ala/acrl/acrlstandards/informationliteracycompetency.htm

Commission of the European Communities. (2002, June 21-22). *E-Europe 2005: An information society for all*. An action plan presented at the Seville European Council, Lisbon (pp. 1-23).

Consulting and Audit Canada. (2004). *Transformation for the digital age: The development of Botswana's national ICT policy. Preliminary report for consultancy for the formation of a national information and communication technology policy technical bid* (pp. 1-81). Toronto: Government of Canada.

Docktor, R. (2004). *Accelerating e-government: Readiness at work*. Retrieved on August 12, 2004, from http://unpan1.un.org/intradoc/groups/public/documents/CAFRAD/UNPAN006617.pdf

Economist Intelligence Unit and IBM Corporation. (2004). *Scandinavia consolidates lead in fifth annual economist intelligence unit e-readiness rankings*. Retrieved on April 14, 2007, from http://www.store.eiu.com/index.asp?layout=pr_storyandpress_id=1260000726 e-Europe+. (2003). *A cooperative effort to implement the information society in Europe: Action plan* (pp. 1-32). Lisbon: European Commission.

Gordon, G. (2003, February 16). SME survey. *Sunday Times Business Times*, 15.

Gray, C., & Lawless, N. (2000). *Innovations in the distance development of SME management skills*. Retrieved on March 4, 2005, from http://www.eurodl.org/materials/contrib/2000/gray.htm

Horrocks, G., & Haines, M. (2004, August 22-27). *Health information literacy and higher education: The King's College London approach.* Paper presented at the World and International Congress: 70th IFLA General Conference and Council, Buenos Aires, Argentina. Retrieved on November 27, 2004, from http://www.ifla.org/IV/ifla70/prog04.htm

Hwang, G. (2004). *Diffusion of information and communication technologies and changes in skills.* Retrieved on August 27, 2004, from http://www.emcc.eurofound.eu.int/content/source/uk02010s.html?p1=sector&p2=Information_Technology

ILO. (2001). *Life at work in the information economy. World Employment Report 2001*. Retrieved on March 11, 2004, from http://www.ilo.org/public/english/support/publ/wer/overview.htm

Lee, C. J., Chi-Keung, L., & Hwang, T. S. (1994). *The APEC survey on small and medium enterprises* (pp. 1-15). Singapore: APEC Secretariat.

Maitlamo. (2004). *ICT sector survey in Botswana. Report of national ICT consultancy for Botswana* (pp. 1-68). Gaborone: Consulting and Audit Canada.

Maksoud, S. S., & Youseff, M. A. A. (2003). *Information and communication technology for small and medium enterprisesin Egypt* (pp. 1-15). Cairo. SME Development Unit, Ministry of Foreign Trade.

McConnell International. (2000). Risk e-business: Seizing the opportunity of global e-readiness. Retrieved on September 14, 2004, from http://www.mcconnellinternational.com/ereadiness/EreadinessReport.htm

Mizzi, J. (2004). *Organisation and cultural transformation: The Malta experience* (pp. 1-93). Blata-Baida: MITTS.

Mutula, S. M., and P. van Brakel (2006). An evaluation of e-readiness assessment tools with respect to information access: Towards an integrated information rich tool. International Journal of Information Management, 26(3), 212–223. doi:10.1016/j.ijinfomgt.2006.02.004

Porter, M. (1990). *The competitive advantage of nations.* New York: The Free Press.

Ramsey, E., P. Ibbotson, J. Bell, and B. Gray (2003). E-Opportunities of service sector SMEs: An Irish cross-border study. Journal of Small Business and Enterprise Development, 10(3), 250–264. doi:10.1108/14626000310489709

SADC e-Readiness Task Force. (2002). *SADC e-readiness review and strategy* (p. 1060). Johannesburg: SADC.

SIDA. (2003). *Information and communication technologies (ICT) in developing countries*. Retrieved on March 11, 2004, from http://www.sida.se/Sida/jsp/polopoly.jsp?d=321

Southwood, R. (2004). The impact of ICT on SMEs-a motor for future economic growth in hard-pressed times. *Balancing Act News Update, 192*. Retrieved on January 26, 2004, from http://www.balancingact-africa.com/news/current1.html

Zakaria, Z. (2001). *Factors related to information technology implementation in the Malaysian Ministry of Education Polytechnics*. Unpublished doctoral dissertation, Faculty of the Virginia Polytechnic Institute and State University, Blacksburgh Virginia.

Section 7
Globalisation, Trends, and Best Practices in the Digital Economy

This section consists of two chapters: Chapters 16 and 17 (Globalisation of the Digital Economy and Trends and Best Practices in the Digital Economy, respectively). Chapter 16 begins with the ICT-driven impact of globalisation on commerce and the competitiveness of SMEs. This is followed by a discussion of global business competitiveness by region-North America, Western Europe, Eastern Europe, Asia-Pacific, Latin America and the Caribbean, the Middle East, and Africa. The opportunities the digital economy engenders for SMEs are then followed by the role policy makers and managers of SMEs can play in this respect. The chapter aims to provide best practices from around the world from which SMEs, especially those in the developing world, can learn.

Chapter 17 expands on the previous chapter by focusing on the global initiatives of governments and private agencies to enhance international trade, especially with regard to e-commerce. The initiatives are categorised by continent-North America (led by the United States and Canada), Europe (spearheaded by the European Union and individual countries such as the UK and Germany), Nordic countries (Norway, Denmark, and Finland), and Asia (South Korea, Mongolia, Taiwan, New Zealand, Hong Kong, China, India, and Japan). Africa does not yet offer much in terms of examples of best practices in the digital environment. The chapter moves on to discuss the concept of a learning organisation in the digital economy.

Chapter 16
Globalisation of the Digital Economy

INTRODUCTION

Data transmitted over the Internet does not recognize national borders. It is this factor that has led information technology to open up new avenues in global commerce. People in separate countries can work on the same project without having to physically relocate. Organizations can now deploy their resources and operations anywhere around the world. Information about new products, corporate earnings, etc; can be shared simultaneously in a networked economy via corporate e-mail systems, value-added networks or over the Internet. In an increasingly globalized business world, for which the Internet is partly responsible, countries are likely to experience the migration of skilled workers to higher paying economies or jobs. Therefore, as the globalization of the world's economy takes shape, concerted efforts must be

DOI: 10.4018/978-1-60566-420-0.ch016

made by national governments and states to train enough students and workers to meet the new challenges brought about by the digital economy (Selhofer, 2003).

As a construct, globalization has made it possible for business enterprises to conduct business operations that extend beyond their areas of fixed abode. It has become increasingly easy for an SME to participate in the world market because of IT, which virtually any business can access. Enhanced access to information by business enterprises implies the ability to reach a greater market audience. This is necessitated even more by the globalization of human capital, characterized by the easy migration of skills between countries. These factors make transacting business in a virtual set-up a profitable reality (Tetuju, 2001). Because of the opportunities offered by information and communication technologies, the vision of perfect competition is also becoming a reality. Competition is fostered by the increasing size of the market opened up through globalization. Consumers can now find out what prices are offered by all the vendors for any product; and new markets and businesses can deliver their products down a phone line anywhere in the world, twenty-four hours a day (Romer, 1990). In the globalized business environment, knowledge spreads fast, and for a firm to be competitive it must be able to innovate more quickly than its competitors. Brands, as products of globalisation, are becoming critical in the global marketplace because they strengthen consumer trust in nations and their products. This is especially true in the current marketplace where consumers are overwhelmed by choice. Like intellectual capital, a brand accounts for a significant proportion of a company's value (Ministry of Economic Development-Canada, 2003).

By leveraging ICTs, SMEs are able to enter global markets that were once only the preserve of large scale companies. However, this also means that the world is growing increasingly competitive, and SMEs have to struggle to survive in an environment where they are surrounded by well resourced enterprises. In such an environment, the Internet offers the potential for improved marketing and benefits from lowered production costs. The Internet can be used by SMEs as a boundless, up-to-the-minute source of information about markets, especially on pricing, overseas trends, products and research. The Net also enables SMEs to actively identify customers and suppliers, and facilitates selling directly to the customer, thus removing the middleman and reducing costs through savings on commission. ICTs in general and the Internet in particular also impact positively on purchasing and production because of how they improve the capacity to purchase goods and services, banking and investment, and access to expert assistance. These benefits, coupled with improved access to information on the latest developments, allow for an increase in the efficiency of production. The Internet further enables the automated tracking of deliveries and provides sophisticated customer relationship management software which is able to capture transaction data and create a database containing information about each customer, such as customer payment preferences, and/or preferred delivery methods (Locke, 2001).

The pervasiveness of ICT in the globalized digital economy, as already demonstrated in the previous chapters, can enable SMEs to benefit further from applications such as client billing, word processing and generally improved business performance. However, the positive impact of ICTs is dependent on a number of factors (Widdison, 1993:5-8). For example, in the manufacturing sector, studies have shown that success in the application of ICTs in SMEs is associated with user participation, the number of administrative applications in the firm, the influence of personality, and the commitment and participation of top management (Montazemi, 1988:241-3).

Increasingly, it has become evident that the competitiveness of (big and small-sized) business enterprises around the world is closely related to their preparedness (e-readiness) to leverage ICTs in order to effectively apply information in their decision making processes. The deployment of ICTs is critical to improved productivity and business competitiveness. For example, it is possible for an artisan in a rural village to use the community centre's computer to sell handicrafts on the World Wide Web, or for a farmer to use a wireless handheld device to research market prices (Sachs, 2003). Anecdotal reports reveal that people are selling their wares on websites, SMEs are providing transcription services via the Internet, and even rural farmers are checking product prices via the Web in places such as India, Kenya, and Uganda.

Business enterprises that are using the Internet to improve their competitiveness in the global market are largely dependent on ICTs. Countries that are leaders in e-readiness also have a competitive edge over those whose levels of e-readiness are low. For example, countries such as Denmark, the United Kingdom, Sweden, Norway, Finland and the United States that are consistently placed in the top tier of e-readiness surveys/rankings also have competitive digital business environments (Economist Intelligence Unit limited and IBM Corporation, 2004:26). The potential of the Internet and its associated technologies to enable global e-commerce has been widely documented (Lee and Clark, 1997:113; Montealegre, 1999:207; Weingarten, 1994:17). Internet-based market structures and the extension of global telecommunication networks offer firms in developing countries new exchange mechanisms that enable them to compete on more equal footing in world markets (Moodley, 2001:90).

Globally, ICTs have changed the nature of global relationships, sources of competitive advantage, and opportunities for economic and social development. ICTs have also changed the way in which businesses interact with their primary stakeholders, i.e. suppliers, customers and investors. Increasingly, success in the information age depends on the widespread integration of ICTs in society at large (Economist Intelligence Unit, 2003). As the Internet continues to pervade all aspects of the social fabric of countries, large (private) firms, small-sized enterprises, and public sector organisations are all increasingly changing their business models to take advantage of the access to information that it (the Internet) provides.

GLOBAL BUSINESS COMPETITIVENESS

Although ICTs are playing a critical role in enabling enterprises to enter the international market and extend their business reach, the degree to which this occurs varies greatly within and especially between developed and developing countries, as described below.

a) North America

The United States and Canada are among the top countries in the world in terms of e-readiness. In both countries, the telecommunications infrastructure is strong, and businesses and consumers have swiftly and effectively integrated online processes into their daily activities. The e-commerce industry in the United States is the world's largest and oldest. 2003's estimates placed the United States' total value of online retail sales at nearly US$ 50 billion (Economist Intelligence Unit and IBM Corporation, 2004:11). North American countries also appreciate the importance of attaining a high level of e-readiness among SMEs. US' SMEs have cited various reasons for using ICTs, such as sending requests for information on products, browsing through lists of products, sending purchase orders, and viewing retail prices (Agriculture and Food Canada, 2001). Other than B2C (business-to-customer) interaction, Internet commerce in North America occurs through business-to-business collaboration in areas such as product design and marketing. In 2002, 20% of Canadian businesses used Internet technologies to interact with other businesses and to develop or test products (Charles et al., 2002).

In a study by Agriculture and Food Canada (2001), SMEs cited various reasons for sourcing online products and services, including: less turn-around time, faster deliveries, enhanced product selection, increased international competitiveness, increased convenience for customers, reduced procurement costs, decreased average transaction costs, efficient purchasing processes, improved cycle time and productivity, decreased inventories, enhanced efficiency and profitability, increased depth of communication, and enhanced open standards. Furthermore, SMEs in Canada's agricultural sector benefited in their e-commerce transactions from the following: the ease with which they could locate hard-to-find products; the provision of electronic catalogues; online advertising; attracting new customers; faster payments; increased sales; increased loyalty of customers; enhanced organisation and production in real time, in any location; and the ability to telework from home.

As already pointed out in the previous chapter, the Canadian E-Business Initiative (CeBI), which was established in 2002 by the Canadian government, is responsible for developing the country's national innovation strategy, and is aimed at addressing the needs and opportunities of the SME sector, which is the backbone of the Canadian economy. The CeBI is also responsible for enhancing research and development (R&D), which are considered important in Canada's quest to become an innovative nation. The CeBI recommends that government initiatives should aim to address the barriers facing SMEs and encourage more sophisticated use of advanced e-business applications for those already connected. The CeBI recommends that the government of Canada, the business community and universities, should continue to work together to create greater opportunities for graduates to gain experience through internships or work placements. The initiative has successfully launched the eCorps program in partnership with Career Edge, which places interns for four months in SMEs across Canada. This pilot program is expected to provide SMEs with access to skilled people while showing young professionals that opportunities exist in the Canadian market. One of the CeBi's requisite recommendations for incentives is based on the belief that the right business and regulatory environment is one that facilitates and encourages innovation. Through the recommendations of the CeBI, Canada has made significant progress in supporting the venture capital (VC) sector and has joined world leaders by coming second in the OECD in venture capital investment (Canadian E-business Initiative, 2003).

b) Western Europe and the Nordic Region

The European Commission is the main body behind various initiatives in Europe aimed at leveraging the use of ICTs to enhance competitive advantage in various sectors amongst the member states. The EU believes that technological developments (e.g. broadband access) in the information society are largely untapped in their potential to bring about significant economic and social benefits. New services, applications and content create new markets and provide the means by which to increase productivity and enhance growth and employment throughout the EU countries' economies. Furthermore, the EU recognises the importance of member countries cooperating in support of e-business in Europe in order to promote the take up of e-business (Commission of the European Communities, 2002:2).

In 2004, most of Western Europe's countries were ranked among the top 25 in the world in terms of their e-readiness statuses. In the same year, the e-readiness global ranking of countries in the Nordic countries showed a highly accelerated penetration of high speed Internet infrastructure and advanced e-business services. For example, Switzerland scored above average (in Europe) in the connectivity, business environment and supporting e-services categories. These categories generally reflect the importance that e-commerce is playing in the global service industries on which the economy depends. (Economist Intelligence Unit and IBM Corporation, 2004:2, 8).

In general, European enterprises were reportedly employing networking technology and e-services to improve operational efficiency in order to service global customers and subsidiaries. In Germany, more than one in three companies was found to be highly dependent on the Internet in their day-to-day operations (Economist Intelligence Unit and IBM Corporation, 2004:9). The positive impact of e-readiness on the operation of SMEs in Western Europe has been apparent. Businesses have been benefiting from the transition to e-commerce because of general increases in productivity; a streamlined interface between production and sales; better stock management; significant reduction in distribution costs, especially for products that can be digitised; reduction in the costs of after-sales service; enhanced advertising and selling in the global market at a lower cost; and an increased turnover in the telecommunications sector as a direct result of the proliferation of e-commerce (eEurope+, 2003:1-10).

The EU has in place an Electronic Commerce Initiative through which it has incorporated e-commerce and information society policies. Following the Lisbon summit in March 2000, the strategic objective endorsed by the European Commission was for the region to become the most competitive and dynamic knowledge-based economy in the world within the next ten years (CEC, 2000).

The British government, in its endeavour to enhance business competitiveness, provides SMEs with specific information through its Small Business Services (SBS) website. This website contains direct links to sources that are relevant to SMEs. The government portal provides an explanation of what e-commerce and e-business can mean to SMEs, and offers advice to all companies, ranging from those currently using ICT, to specialist multimedia hardware and software developers. There are free business publications available on the site on various technologies to explain IT jargon, which can be alienating to non-technical people. The British government has also established *UK Online for Business* - a network of advisers who form part of Business Link in England and their counterparts in Scotland, Wales, and Northern Ireland. They are expected to provide advice suited to local companies' needs, including e-commerce services ranging from direct company-specific assistance, to training and consultancy (Adeshara et al. 2004). The presence of extensive information that is aimed specifically at SMEs on adopting ICT and e-commerce highlights the UK government's commitment to promoting partnerships between business, government and the wider community. This is further underpinned by the complete modernisation of the country's government. The emphasis is on encouraging SMEs' success in implementing ICT and competing both locally and internationally.

c) Eastern Europe

In Eastern Europe, there has been a gradual market liberalisation of the telecommunications sector, allowing competition to set in and leading to the consequent improvement of services and the lowering of prices. Countries such as Estonia, the Czech Republic, Hungary, Slovenia, Latvia, Poland, Lithuania and Slovakia, have achieved decent infrastructure and e-business environments. For instance, in 2004, all of Estonia's schools had broadband connection and 80% of the country's banking transactions were electronic (Economist Intelligence Unit and IBM Corporation, 2004:18). The region is leveraging its proximity to Western Europe to create niches in software development and high-tech manufacturing, and outsourcing services that cater for Western European businesses. The whole of Eastern Europe is leveraging mobile communications for electronic business transactions. For example, in 2004, Russia was ranked by the Economist Intelligence Unit and IBM Corporation to be among the world's fastest growing mobile markets, while the Czech Republic and Hungary had mobile phone penetration of 84% and 74% respectively. The high prevalence of mobile communications in these countries could have stemmed from growing competition amongst domestic and international players, resulting in lower prices. According to 2004's e-readiness ranking, countries such as the Czech Republic (28th), Hungary (28th) and Slovakia (39th) had skilfully transformed their technology manufacturing industries into a global source of IT service support and Internet-enabled customer care (Economist Intelligence Unit and IBM Corporation, 2004:4).

d) Asia-Pacific

Most of the top countries in 2004's e-readiness ranking within the Asia-Pacific region were also leaders in business competitiveness. These countries included Singapore, Hong Kong, Australia, South Korea, New Zealand, Taiwan and Japan. Singapore and Hong Kong have rapidly growing e-trading platforms, especially in the regional finance market. Elsewhere in the region, Malaysia has enjoyed significant economic growth over the years due to the country's export-based economy, which is largely dominated by ICT-related growth with good foreign investment from Japan and the

United States. South Korea, which was one of the leading countries in the broadband market with 27% in terms of Internet penetration in 2004 e-readiness survey, has a thriving e-tailing and online gaming sector. Moreover, cheap and abundant bandwidth for homes and small-sized enterprises drove the uptake of e-business tools and the growth in online transactions within the country (Maitlamo, 2004:10).

Likewise, India and China have growing e-business and e-enabled service sectors. This most probably led to their strong standing in the e-readiness ranking in 2004, at 46th and 52nd positions respectively. India's thriving software and outsourcing industries were estimated to contribute 3% to the country's GDP in 2004. Furthermore, India's famed IT-enabled services sector was estimated to be contributing US$ 17 billion to the economy annually. The application of ICTs in business enterprises across the country has made quality information available to individuals; reduced the digital divide between firms and the outside world; created new business opportunities; maintained interconnectivity across nations; assisted with overcoming virtual and physical isolation; and reduced transaction costs and time. India's success is being replicated in regional countries such as the Philippines, Malaysia, Korea, Japan and China, where booming call centres, customer help desks, and software production houses have emerged (Economist Intelligence Unit and IBM Corporation, 2004:4, 15). In the Philippines, the deployment of new technologies in different sectors of society in 2001 made enterprises within the region much more inter-dependent by accelerating the movement of goods, services, ideas and capital across borders (Zakaria, 2001).

e) Latin America/The Caribbean

South American governments have recognised the economic value of moving their business processes online in order to cut costs and improve transparency. Generally, governments in the region are creating policies and legislations that support e-business development. For example, by 2004, Mexico had established a law on digital signatures; companies in Chile were eligible to start electronic invoicing; and in Peru, the government was looking into building an e-business platform for its shipping and transport industries. In Columbia, the growth in mobile communications has driven prices down, with mobile subscriptions significantly exceeding subscriptions to land-lines (Economist Intelligence Unit and IBM Corporation, 2004:12).

The Economist Intelligence Unit's 2004 global e-readiness ranking revealed that Latin American outsourcing powerhouses, such as Brazil (35th), Argentina (37th) and Mexico (40th), had created a niche for themselves in software technology development. Brazil, for example, has a thriving IT-enabled services sector that grew from a combination of efficient back-office facilities, and competent, cost-efficient workers. Within the Caribbean, an estimated 5000 women were employed in data processing activities in the late 90s as a direct result of the introduction of ICTs to the SME sector. Such a high employment of women, who are often marginalised in many countries, demonstrates to other developing countries how an affirmative stance in global markets may be enhanced.

f) The Middle East and Africa

By 2004, the Middle East and Africa had not yet significantly taken advantage of ICTs to enhance their business competitiveness, in part because they suffered low e-readiness statuses. The two regions, particularly the Middle East, also suffered from strict control over Internet content and the over-regulation of service providers by their governments. Consequently, both regions had poor infrastructure and rather unwelcoming business environments. For example in Algeria, which was ranked [61st] globally in terms of its e-readiness status in 2004, e-commerce continued to be hindered by poor telecommunications infrastructure. During that year, Algeria had 16 telephone users per 1,000 people. Comparatively, Saudi Arabia's e-commerce had yet to emerge because of a poor telecommunications grid, slow Internet access and high service prices. However, small steps towards lowering prices have since enticed some Saudi companies to adopt e-business practices (Economist Intelligence Unit and IBM Corporation, 2004:18).

Despite the low status of e-readiness in the Middle East and Africa, there are some countries within the two regions that have made good progress in enhancing their e-business environments. For example, the South African government spends approximately US$ 12 billion annually on its own IT infrastructure to support e-government interfaces and processes, such as revenue services, e-filing, and electronic tax filing [the latter allow citizens and businesses nationwide to submit tax returns] (Economist Intelligence Unit and IBM Corporation, 2004:16). Governments in the two regions have also recognised the importance of ICTs as a competitive tool in both regional and global markets. For example, e-Africa Commission's (NEPAD's) ICT Task Team was founded in 2002 and charged with the responsibility of developing ICT programs for the region to promote their competitive positioning in global e-markets. The additional responsibilities of the e-Africa Commission include serving as the principal advisory body to NEPAD's Heads of State and the Government Implementation Committee on ICT matters; developing a NEPAD strategy and action plan for the accelerated development of ICT infrastructure, services, applications and content to meet Africa's development needs; and promoting the use of ICTs in support of other NEPAD programmes (Chasia, 2002:6).

The SADC World Economic Forum meeting of the Heads of State (held in

Johannesburg in 2001) recognised that ICTs promised enormous economic and social benefits because they instil independence in people. They felt that developing countries in general and the SADC member countries in particular faced the threat of being left behind if they did not address the growing digital divide between them and developed countries, and within their own countries. Among other recommendations, the forum underlined the need for the region to develop e-commerce infrastructure in order to enable individual member countries to participate in global trade, enhance trade on a regional basis, and facilitate interaction between the SADC member states; with a special emphasis on extending these benefits to SMEs (SADC E-readiness Task Force, 2002:6).

Within individual SADC member countries, the competitive business advantages of deploying ICTs are well recognised. For example in Mauritius, there has been a successful shift from an agriculture-based economy, to a more diversified economy involving foreign investment in IT outsourcing industries aimed at tapping business from India and South Africa. Likewise, it is worth noting that even though Namibia's economy is heavily dependent on mineral exports, the country has started to seek out new foreign investment opportunities, including ICT growth in the region (Maitlamo, 2004:10).

OPPORTUNITIES FOR SMEs IN THE ECONOMY

The potential of SMEs to enhance their business competition using ICTs in any economy has already been discussed in the preceding chapters. The digital economy affords SMEs the benefits of (Rizk, 2004):

- Reduced transaction costs
- Removal of barriers to market entry
- Opportunities to gain a competitive advantage over larger firms due to economies of scale
- Using and managing supply chain networks to facilitate global business reach
- Tools for marketing and distribution. These facilitate responsiveness to market demand and the customisation of offerings
- Opportunities for innovation and the emergence of new products and services.

ICTs offer SMEs the potential to increase exports, promote growth, and encourage human development. Within the digital economy, the e-readiness of enterprises ensures 24/7 access to information, thus contributing to their efficiency. OECD (2000) noted that ICTs help small-sized business entrepreneurs overcome information poverty, which makes them more connected, more certain, less risk averse and more capable of informed decision making.

New technologies, especially Internet-related applications, are having a significant impact on the operations of SMEs, and are arguably essential for the future survival and growth of nations' economies in general and SMEs in particular. Through the use of relevant ICTs, SMEs are able to benefit by: selling their products and services to international customers; developing their ability to manage information infrastructure for business; improving their use of external information intensive resources; improving their capacity to efficiently manage information on a global scale; and gaining access to the free and rapid flow of information (Ramsey et al., 2003:253; Minton, 2003).

Internet-based market structures and the extension of global telecommunications networks appear to offer production firms in developing countries new exchange mechanisms to enhance their foothold in global business environments. Business-to-business e-commerce, for example, has the potential to reduce information asymmetries and trade–related transaction costs, provide the firm with an international profile, and eliminate market barriers inhibiting the growth of developing countries' exports. For example, small-sized businesses in the South African wood and furniture manufacturing sector apply the Internet to collaborate with trading partners online, develop close information-based links with suppliers, and enhance access to information. In most developing countries, it is important to use the Internet to trade in the international market because the small-size of local markets makes it difficult to sustain economies of scale (Moodley, 2001:90). The government of Botswana (Republic of Botswana, 2003), for example, recognises that the use of technology to support SMEs would enable them to gain competitive advantage by creating value, extending their market reach, and creating a more level playing field for competition with large firms.

As the world economy continues to move towards increased integration as a result of advances in ICT and diminishing trade barriers, some of the greatest opportunities for small-sized businesses will be derived from their ability to participate in regional and international markets through new technologies such as computing, communications and multimedia technologies, which are changing the way we work and learn. These technologies are changing the global flows of information, trade and investment, and the competitive advantage of industries, services, countries and regions. Increasingly, e-commerce presents an opportunity to compensate for the traditional weakness of SMEs in areas such as access to new markets and gathering and distributing information on a broad and international scale. ICTs are being used to create dynamic business strategies in a paradigm, shifting environment that raises customers' expectations about speed, comparability, and prices (Ramsey et al., 2003:250).

SMEs are increasingly being recognised by many countries as the glue that holds many economies together. Within the Southern African Development Community (SADC), SME policy initiatives now rank highly on the political agenda of eradicating poverty in the region (Chidzomba, 2002). SADC E-readiness Task Force (2002:6) underscores the importance of empowering SMEs through the use of ICT in order for them to participate in global trade. The SADC member states also recognise that SMEs play a critical role as a source of most new forms of employment. Consequently, collaborative efforts were initiated in 1996 by member states to enhance the growth of the SME sector. This culminated in the formation of the Small Enterprise Promotion Advisory Council (SEPAC). SEPAC serves as a regional SME support network for the SADC region, and is responsible for promoting the sustainable regional integration of SMEs in SADC; marketing, business linkages and cross-border trade; entrepreneurship development and training; and providing access to finance. SEPAC also serves to enable information technology development and transfer; facilitate the exchange of information and experiences; identify specific barriers to small-sized enterprises' success; support policies and programmes that provide small-sized businesses with increased opportunities; and enhance the national, regional and international performance of regional SMEs by reducing impediments to growth.

Many developing countries face the challenges of unemployment and poverty, and for this reason, governments and donor agencies have been emphasising the role that the SME sector can play in promoting economic and social development. This has been followed by the channelling of a lot of financial assistance into initiatives that encourage and empower SMEs. Some governments are investing in SMEs as a way to diversify their economies. Botswana's government, for example, has realised that it is not sensible to depend on mining for the long-term economic development of the country, and is now encouraging the development of the SME sector in order to diversify its economy (Machacha, 2002:277; Amani and Mbaga-Kida, 2001).

POLICY AND MANAGERIAL ROLES IN ENHANCING THE COMPETITIVENESS OF SMEs

In order for SMEs to comfortably partake in the international business environment, a number of policy and managerial interventions are necessary. Governments, especially those in the developing world, should put in place facilitation mechanisms that enable SMEs to gain access to credit and finance from commercial banks. It is also important for the governments to identify various business models that SMEs can afford, manage and maintain. Action is needed from the government to veto the tax on computers and related products to push the prices down in order to enhance the uptake of ICTs by SMEs. Consideration should be given to a funding mechanism, such as an (SME) access fund, to enhance SMEs' access to ICTs and also promote connectivity. Governments may also wish to provide grants, subsidies, corporate donations, tax holidays or exemptions, and special zones where regulatory or pricing constraints are more flexible.

In countries where they do not exist, the respective governments should consider setting up an export-import bank and other small-sized business administration structures that would offer pre-export financial support for SMEs in the form of loans. Governments should also consider relaxing tariffs in order to improve the free flow of both trade and services within the SME sector. Furthermore, developing private venture capital markets and offering credit guarantee facilities to firms with international potential would assist SMEs in the international business environment. Instilling a trade policy that gives incentives to SMEs would also enable them to compete more effectively against bigger companies. A simplified industrial licensing system is necessary to reduce bureaucracy and speed up the issuance of licenses and bring services closer to the businesses. Where possible, governments should put in place a Small Solutions Centre for SMEs. The Centre would support the SMEs by providing reliable market information, access to finance and capacity building. Such centres already exist in Kenya and Madagascar.

There is a need for strong education and support programs for SMEs in order to encourage them to do business in foreign markets. Governments should also facilitate and enable SMEs to identify and work with customers and suppliers across borders. Awareness should be created amongst the SMEs about the benefits of forming partnerships with multinational companies and the potential of e-commerce in order to enhance their export capacity. Consideration should also be given to working through bilateral and multilateral channels to foster global partnerships that would promote the SMEs' development. In order to ensure that the products being delivered are of a high international quality, there is a need for standardisation to ensure that they (the products and services) are consistent, compatible, safe and effective. Governments can also enhance competitiveness in their SME sectors by improving network performance through various regulatory steps that end with effective ICT and e-commerce infrastructure in place.

Through research and development, efforts should be made to explore innovative technological solutions that must be affordable for use by small-sized businesses, and which they could sustain within local economic conditions. Likewise, efforts should be made to provide applications, content, and services that are relevant to their businesses. Finally, governments should support the private sector and encourage investment in SMEs by reducing bureaucratic red tape and retraining public officers to reduce unnecessary procedures in the licensing of businesses in the country and thereby enhance efficiency. For these recommendations to be achieved, an overarching policy and legislative framework that underpins access to and the application of information by SMEs to their operations must be developed.

CONCLUSION

Businesses, both large and small-sized enterprises are leveraging the opportunities provided by ICT to extend their market reach within the global business environment. The Internet has increased business' trans-border data flow and facilitated access to market intelligence by SMEs. With the increasing globalisation of the business environment and the easy movement of human capital across borders, it is now possible for SMEs to mobilise resources and operations anywhere in the world by leveraging ICT.

Countries that are effectively leveraging ICTs for business advantage, especially in the developed world, have high levels of e-readiness. Such countries are finding it much easier for their companies to successfully penetrate and operate in the global business environment than enterprises in developing countries where the level of e-readiness is low. It is anticipated that with the globalisation of the digital economy, SMEs will find it prudent to enter the international market because they stand to benefit more from gaining competitive advantages and making use of management supply chain networks. Enterprises aspiring to take advantage of the digital economy by leveraging ICTs can learn from the developed world, where most enterprises, by virtue of the high levels of e-readiness of their countries, are already partaking in the global digital economy.

REFERENCES

Adeshara, P., R. Juric, J. Kuljis, and R. Paul (2004). A survey of acceptance of e-government services in the UK. Journal of Computing and Information Technology, 12(2), 143–150. doi:10.2498/cit.2004.02.10

Agriculture and Food Canada. (2001). *The e-readiness of Canadian Agriculture and Food Canada SMEs*. Retrieved on March 21, 2004, from http://www.agr.gc.ca/misb/fb/food/ecommerce/eready/introduction_e.html

Amani, H. K. R., & Mbaga-Kida, T. (2001). The role of human capabilities in economic development. *African Journal of Finance and Management, 10*(1). Retrieved on August 11, 2004, from http://www.inasp.info/ajol/journals/ajfm/vol10no1abs.html

CEC. (2000). Strategic objectives 2000-2005: Shaping the new Europe. *COM,154.*

Canadian E-business Initiative. (2003). *Submission—the Canadian e-business initiative.* Retrieved October 24, 2008, from http://www.cebi.ca/

Charles, S., Ivis, M., & Leduc, A. (2002). Embracing e- businesses: Does size matter? *Connectiveness Series.*

Chasia, H. (2002, November 14-15). *The NEPAD ICT programme.* Paper presented at the NEPAD Communications and Outreach Workshop, Sandton, South Africa (pp. 1-29).

Chidzomba, J. (2002, August 7-10). *Experience on regional networking of SMEs in the case of small enterprise promotion advisory council [SEPAC].* Paper presented at Seminar on Promotion of SMEs in Angola –Experiences & Lessons Learnt, Luanda, Angola. Retrieved on August 11, 2004, from http://spesmru.intnet.mu/sepac/luanda_paper.html

Commission of the European Communities. (2002, June 21-22). *e-Europe 2005: An information society for all.* An action plan presented at the Seville European Council, Lisbon (pp. 1-23).

Department of IT eTechnology Group (India). (2003). *Assessment of central ministries and departments. E-Governance readiness assessment 2003 draft report, 48.*

Economist Intelligence Unit and IBM Corporation. (2004). *Scandinavia consolidates lead in fifth annual economist intelligence unit e-readiness rankings.* Retrieved on April 14, 2007, from http://www.store.eiu.com/index.asp?layout=pr_storyandpress_id=1260000726 e-Europe+. (2003). *A cooperative effort to implement the information society in Europe: Action plan* (pp. 1-32). Lisbon: European Commission.

Lee, H. G., and T. H. Clark (1997). Market process reengineering through electronic market systems: Opportunities and challenges. Journal of Management Information Systems, 13(3), 113–136.

Machacha, L. (2002, April 3-6). Impact of information technology on small and medium enterprises (SMEs) in Botswana. In *Proceedings of International Conference, Port Elizabeth, South Africa* (pp. 277-282).

Maitlamo. (2004). *ICT sector survey in Botswana. Report of national ICT consultancy for Botswana* (pp. 1-68). Gaborone: Consulting and Audit Canada.

Maksoud, S. S., & Youseff, M. A. A. (2003). *Information and communication technology for small and medium enterprises in Egypt* (pp. 1-15). Cairo: SME Development Unit, Ministry of Foreign Trade.

Ministry of Economic Development-Canada. (2003). *Bank lending practices to small and medium sized enterprises*. Retrieved on March 5, 2005, from http://www.med.govt.nz/irdev/ind_dev/access-to-finance/lending-practices/lending-practices-06.html

Minton, S. (2003). *Nordic nations still top information index*. Retrieved on August 13, 2004, from http://www.worldpaper.com/2003/oct02/isi1.html

Montazemi, A. R. (1988). Factors affecting information satisfaction in the context of the small business environment. MIS Quarterly, ▪▪▪, 239–256. doi:10.2307/248849

Montealegre, R. (1999). A temporal model of institutional interventions for information technology adoption in less developed countries. Journal of Management Systems, 16(1), 207–240.

Moodley, S. (2001). Impact of electronic commerce and small exporting firms in the South African wooden furniture manufacturing sector. Journal of Information Technology Impact, 2(3), 89–104.

OECD. (2000, June 14-15). Realising the potential of electronic commerce for SMEs in the global economy. *Conference for Ministers Responsible for SMEs and Industry Ministers,* Bologna: Italy. Retrieved on August 8, 2004, from http://www.conferenzabologna.ipi.it/Eng/dopo_la_conferenza/documenti/WK3_eng.pdf

Ramsey, E., P. Ibbotson, J. Bell, and B. Gray (2003). E-Opportunities of service sector SMEs: An Irish cross-border study. Journal of Small Business and Enterprise Development, 10(3), 250–264. doi:10.1108/14626000310489709

Republic of Botswana. (2003). *Budget speech 2003* (pp. 1-39). Gaborone: Government Printer.

Rizk, N. (2004). *E-readiness assessment of small and medium enterprises in Egypt: A micro study.* Cairo: American University. Retrieved on August 9, 2004, from http://www.sba.luc.edu/orgs/meea/volume6/Rizk.htm

Romer, P. M. (1990). Endogenous technical change. *Journal of Political Econom,* 71-103.

Sachs, J. D. (2003). *Readiness for networked world.* Retrieved on November 28, 2003, from http://cyber.law.harvard.edu/readinessguide/intro.html

SADC E-readiness Task Force. (2002). *SADC e-readiness review and strategy* (p. 1060). Johannesburg: SADC.

Selhofer, H. (2003, June 3). *Indicators for the new e-business environment: Challenges for statistics and research.* A presentation at the DEEDs 5th Policy Group Meeting, Brussels.

Tuteja, S. K. (2000, September 28-29). *Small business survival & growth in the context of globalization. The Indian scenario.* Paper presented at the APO Study Meeting on Small Business Survival & Growth, Jakarta, Indonesia.

Weingarten, F. W. (1994). Public interest and NII. Communications of the ACM, 37(3), 17–19. doi:10.1145/175247.175261

Widdison, R. (1993). Databases in law firms: An overview. Law Technology Journal, 13, 5–11.

Zakaria, Z. (2001). *Factors related to information technology implementation in the Malaysian Ministry of Education Polytechnics.* Unpublished doctoral dissertation, Faculty of the Virginia Polytechnic Institute and State University, Blacksburgh Virginia.

Chapter 17
Trends and Best Practices in the Digital Economy

INTRODUCTION

Various global initiatives are taking place in the digital economy that aim to enhance global business trade and development, especially with regard to e-commerce. The International Chamber of Commerce - ICC (1997), for example, has been involved in a few initiatives aimed at promoting global trust in electronic trade transactions by defining, identifying, and deconstructing what constitutes best practices within the digital economy. The International Chamber of Commerce is a non-profit, private international organization that works to promote and support global trade. It also codifies and harmonizes world trade practices in various publications. The ICC focuses on three distinct areas, namely General Usage for International Digitally Ensured Commerce (GUIDEC); the Electronic Trade Practices Working Group; and

DOI: 10.4018/978-1-60566-420-0.ch017

E-terms Service. GUIDEC expects to instil trust in open electronic commerce by developing a common understanding of the mechanisms used to guarantee identities and authenticate transactions over electronic networks. GUIDEC was established to provide guidance in a market where different definitions in different jurisdictions threaten to undermine the utility of digital signatures in cross-border trade.

The Electronic Trade Practices Working Group functions as a set of founding rules for electronic trade and settlement. Its primary objective is to make trade more efficient by not only adapting rules to suit new technologies and media (such as the Internet), but by also taking advantage of these new tools to streamline trade transactions. This mandate is motivated by the fact that presently, most buyers and sellers in different parts of the world have no legal framework to cover how they conduct their negotiations, make contracts, and/or arrange for finance, transport or insurance online. Most of the rules that apply to international trade still presume the use of paper.

E-terms service is based on an online repository that contains the tools necessary to bring the level of legal risk down to a minimum while composing contracts online and/or conducting electronic transactions. Rules and directives of different kinds that may apply to different jurisdictions in the digital environment can be incorporated into electronic contracts by referring to a unique identifier automatically supplied by the e-terms repository. E-terms service can be useful to SMEs that do not have their own in-house legal expertise.

NORTH AMERICAN DIGITAL ECONOMY INITIATIVES

Countries in the developed world have a number of successful e-commerce initiatives that provide examples of best practices to those just starting to implement e-commerce programmes. In North America, both the United States and Canada have achieved high rates of e-readiness, with an e-commerce absorption rate of about 70%, placing them among the world's leaders in e-commerce assessments. Canada has consistently been highly ranked in the global league of nations in e-readiness and e-government assessments. Among the initiatives that have been undertaken by the state, is *Government of Canada Online*, an official government website designed to make interactions with stakeholders, including businesses, a lot easier. The website is characterised by its focus on users, connectedness, tailor-made online services, and the integration of services (Government of Canada, 2006).

Canada has invested several billions of Canadian dollars in creating knowledge and encouraging innovation through various federal-level research and development funding initiatives. For example, every school and library in the country is connected and involved in constant skills' upgrades. Through the community access program, Canada has connected a number of rural communities to the Net, and also provides the International Computer Driving License curriculum for professionals and teachers to upgrade IT skills (Jutla et al. 2002). Furthermore, by promoting research and development as well as knowledge in SMEs, the Canadian government has improved their export potential and capacity to create jobs (Lefebvre and Lefebvre, 2000). The overall economic climate for e-business investment in the country is favourable because an online policy guiding access, standards, and ways of addressing privacy issues, is in place (Government of Canada, 2006). Canada has also recognized the equivalence of e-signatures and e-documents to their written, paper-based counterparts.

The US Federal government, working with the United Nations Commission on International Trade Law (UNCITRAL), developed a model law that supports the commercial use of international contracts in electronic commerce. The government has also been working with the International Chamber of Commerce to issue model commercial code guidelines. The US Federal government believes that no new discriminatory taxes should be imposed on Internet commerce. The state further believes that no customs duties should be imposed on electronic transmissions. They argue that the application of taxation on commerce conducted over the Internet should be consistent with the established principles of international taxation. They also understand that the law should be neutral with respect to other forms of commerce to avoid inconsistent national tax jurisdictions and double taxation (The New Economy Report, 1999).

In 1999, the US Federal government took the lead in tackling issues of protection and privacy online. Today the US government also takes the lead in encouraging the private sector to establish codes of conduct and self-regulation through rules that consumers must first be aware of before registering for compliance. (The rules should make a provision for recourse should injuries result from non-compliance. The government argues that consumers need to know the identity of the collector of their personal information, the intended uses of the information, and the means by which they may limit its disclosure. They should be given the opportunity to exercise choice with respect to whether or not and how their personal information is used. Likewise, companies creating, maintaining, using or disseminating records of identifiable personal data must ensure that it achieves its intended use, and take every precaution to protect it from loss, misuse, alteration, or destruction. A consumer should also have the opportunity to access, in a reasonable and appropriate manner, personal information that has been compiled by a company, and be able to correct or amend that information when necessary (The New Economy Report, 1999).

EUROPEAN DIGITAL ECONOMY INITIATIVES

Within the European Union, member states have taken a number of measures to meet e-commerce requirements. The e-Europe Action Plan, for example, envisaged a universal, dynamic e-business environment in Europe by 2005. Through the Go Digital initiative, e-Business W@tch - a programme measuring the uptake of ICT and e-business in the EU - was developed in order to monitor, analyse and compare the development of businesses in different sectors of the European economy. A benchmarking exercise conducted in 2005 in 10 sectors of the EU economy in 2004/05 showed that significant progress had been made. For example, (EUROPA - Europe's Information Society Thematic Porta, 2007):

- 45% of EU firms (weighed by employment) had introduced new business processes in 2004
- About 75% of this innovation activity was directly related to or enabled by ICT
- Small, medium-sized and large companies had reached different levels of maturity
- Electronic business was reaching technological maturity, aided by the increased penetration of broadband connections to the Internet
- More than 90% of all firms in all sizes and sectors (except for food and beverages) were connected to the Internet
- Almost half of the firms' purchases occurred online
- In B2C electronic commerce, more than a third of the firms sold their services online
- 45% of the firms (through employment) had introduced new or significantly improved internal processes.

Most of the EU member states have reshaped their regulatory environments to facilitate communication networks and services and promote e-commerce, and have opened doors to new generations of mobile and multimedia services (European Commission, 2002). By and large, in all the European member states, Internet penetration has doubled, telecommunication frameworks are in place, Internet prices have fallen, and e-commerce legal frameworks have been established. Efforts being made to enhance e-commerce within the EU member states include attaining universal digital literacy; facilitating access and the effective use of ICT; addressing digital divide issues; enhancing ICT skills and development; facilitating the use of ICTs; and enhancing lifelong learning, particularly for those whose access to these technologies has been or is inadequate (e-Europe Action Plan, 2004). The European

Union has had a substantial influence on the development and implementation of e-commerce in different member states, especially with respect to the funding of collaborative research between industry and universities, in providing broadband access, and in the standardization of messages and technologies. The member states have implemented the region's Electronic Signature Directive, which defines the use of electronic signatures in various member states' laws. The European Union has also passed directives that deal with the taxation of e-commerce (Mann et al., 2000).

Scandinavian countries are the leaders in e-commerce in the Nordic region, and have developed a number of initiatives to promote e-commerce. Examples of such initiatives include: knowledge diffusion, including the dissemination of information; the development of skills and alliances with business associations designed to create a positive dialogue on e-commerce; economic incentives, such as the provision of favourable pricing for network services, tax breaks to help facilitate the purchase of home PCs, and direct government subsidization of EC activities; directives and legislations on deregulation, regulation, privatization, and the liberalization of the telecommunications market; promulgation of the technical standards used in business transactions, such as encryption and certification measures; and e-governance that focuses on the work processes within the public sector's own organizations and on the services provided to the public (Andersen et al., 2003).

In Denmark, educational efforts in the area of IT and e-commerce have been aggressively promoted by the government since 1995. There is also a very well established system of vocational training for employees organised by employers in collaboration with local vocational training centres. Deliberate efforts have been made to enhance knowledge diffusion to increase the awareness and discourse on e-commerce and help stimulate the adoption of electronic data interactions and Internet-based e-commerce, particularly in the B2B segment. Andersen et al. (2003) noted that in 1997, the government launched a bill that allows companies to provide their employees with a tax-free PC at home as long as the employee takes a number of PC courses. Other e-commerce developments in Denmark include: the liberalization of the telecommunications sector, which has increased the number of companies operating in the market and reduced prices; the use of e-procurement to jump-start e-commerce; policy initiatives to provide online services in businesses; the development of standardised digital infrastructure within government to reduce the processing time of business communications; a government directive that only electronic invoices from February, 2005, would be accepted; the establishment of an active e-procurement policy for the public sector; and a very high focus on the digitisation of government.

Finland has made great efforts to ensure that its population is comfortable in the e-commerce environment. The government implemented exhaustive IT education for its entire population and had successfully achieved the education of most of its citizenry by 1997. By 2003, email addresses had been given to every citizen, and free Internet access was available in libraries. Moreover, the PC dissemination rate was particularly high in educational institutions. The national treasury had invested in IT education in business enterprises to promote IT application in all industries. Finland allows people who have been left behind in education to undergo social education as many times as is necessary. In terms of computer security, the country conforms to guidelines and rules that have been established by the European Union (Washisu, 2003).

Access to the digital economy (i.e. access to networks/clusters) appears to be the main policy objective in the UK. Priority is also given to the generation of new knowledge through research. The British government also has an innovation policy that has been formulated in line with free market forces, privatisation and deregulation of national economy. The Department of Trade and Industry (DIT), a key government actor in innovation policy, focuses more on the diffusion of best practices and on the creation of networks and business support services (Bojic, 2001).

In Germany, improving access to codified knowledge and generating new knowledge through the development of ICT infrastructure and networks appear to be the main policy objectives. The main priorities in Germany's innovation policies are (Bojic, 2001): the adoption of tax reforms; technology transfer between industry and science - providing incentives to universities to market new technologies produced in laboratories, stimulating the mobility of personnel between industry and science, etc; and support of SMEs in R&D and innovation. For the government of Greece, access to digital infrastructure and technology coupled with the development of ICT and telecommunications infrastructure and networks are the country's main policy objectives. Other priorities include equal opportunities and access for all; offering better services to citizens and firms; and developing the national communications infrastructure through regulatory reforms in telecommunications (Bojic, 2001).

The government of Norway has put in place the eNorway plan, which aims to promote e-commerce in the commercial business sector and establish Norway as a leader in electronic public administration. The eNorway action plan also focuses on improving broadband Internet connections and access, having recognized that broadband access is an important enabler for large-scale e-commerce adoption. Electronic signatures are fully acknowledged in Norway as a legal form of verification. This move is important in securing e-commerce transactions. The eNorway action plan has helped the country's SME sector by (Bojic, 2001):

- Promoting e-commerce and new venture start-ups among SMEs
- Providing promotional and educational services for the e-commerce initiatives of SMEs
- Creating awareness and increasing knowledge about e-commerce among SMEs
- Helping SMEs' management develop and formulate e-strategies and develop e-action plans for implementation
- Making e-commerce more feasible by analysing pertinent problem areas and promoting pilot programmes
- Providing web portal services that include comprehensive guides and answers to questions regarding e-commerce technology, law, research, markets, etc
- Providing an online strategy guide for the use of SME management.

Through eNorway, SMEs are also benefiting from: e-commerce consultancies; e-commerce workshops for e-strategy implementation; a comprehensive e-commerce resource centre; and the financing of e-commerce projects (Bojic, 2001).

ASIAN DIGITAL ECONOMY INITIATIVES

In Asia, South Korea has made significant progress in e-commerce. The government of South Korea supports programs for B2B and IT use in SMEs. The government has promulgated an e-business National Vision that applies to various governmental organizations, and is also implementing e-commerce law and human resources development. With regard to the globalization of e-commerce, South Korea is actively involved in the full-scale promotion of e-trade through the Pan-Asian E-commerce Alliance with Japan and Germany. The country is also involved in a Japan-Korea e-Trade hub development project. The Korean government has actively engaged 3 million SME managers in e-business in order to strengthen and promote IT in the sector (Kim, 2007).

Although e-commerce in China was still in its infancy in 2000 in terms of technology and market saturation, interest towards e-commerce was high, with most enterprises expectations highly in favour of B2B commerce. For example, China Mobile, a cellular phone service provider, had introduced a form of e-commerce on mobile telephones where payments were collected by adding on to telephone charges. Many e-commerce sites already existed, key among which were shopping channels, online malls, and a generally large online community of international traders, buyers and manufacturers (Fan, 2000). In nearby Singapore, the Intelligent Island initiative launched a project known as *SingaporeOne*, which enabled broadband Internet infrastructure to be put in place across the entire country. This was followed by the launch of the Electronic Commerce Hotbed programme, the Electronic Commerce Co-ordination Committee (EC3), and the prioritisation of the security of e-transactions (Fan, 2000).

In Mongolia, an ICT Committee chaired by the Prime Minister, with representatives from government agencies, science organizations, and the business community, acts as the main driving force behind the country's e-commerce program. In 1997, the government successfully established a B2B e-commerce system that supplies consumer goods to people in rural areas and raw materials to the industry. The project focused on trade-data exchange between suppliers and buyers, and on developing an e-commerce environment that would facilitate the country's entry into the international network. The project established an inter-media oriented market through B2B e-commerce service centres that support e-commerce trading actions. Import taxes for computer and software goods within the country are low because of the ICT policies of the government. Hence, purchasing IT products in the country is relatively easy and affordable. Most small-sized businesses have at least one computer, which citizens mainly use for word processing and spreadsheet purposes, or in retail-related businesses as cheap point-of-sale terminals. Mid- to large-scale businesses have one computer per employee and servers that run e-mail systems and departmental databases (ICTA of Mongolia, 2006).

Taiwan joins the ranks of global leaders in e-adoption. E-commerce plays a very important role in the country in improving the capabilities of individual enterprises, promoting collaboration and integration between enterprises, and enhancing the competitiveness of the Taiwanese industry as a whole. The Ministry of Economic Affairs, through the Department of Industrial Technology (DOIT), implemented the long-term monitoring of international trends and domestic needs for e-commerce. Information obtained from international and domestic markets has been used to formulate strategies and policies for the development of industrial technology in Taiwan. DOIT has collaborated with other government agencies to ensure that the technologies developed are adopted by the Taiwanese industry and used effectively (Find, 2004).

Hong Kong offers a favorable environment for the development of electronic commerce through the Internet. A company's profits from e-commerce are not subject to tax as long as its business operations are located outside the country's jurisdiction. This is the case regardless of whether or not the company's Internet service provider is based in Hong Kong. In 2005, 1.8% of Hong Kong's companies had sold goods, services or information electronically, which was up by 0.6% from 2004. Through the "Digital 21" IT Strategy, the government is committed to transforming

Hong Kong into the Asia Pacific hub for both Internet traffic and content. Since the emergence of the first commercial Internet Service Providers (ISPs) in 1993, Internet use among Hong Kong's businesses and households increased exponentially, to the point that 87% of the country's households had personal computers (PCs) by September 2006, and 66.9% of the household computers were connected to the Internet in a population of about 7 million. In 2005, 54.7% of the businesses within the country had Internet connectivity. And as at September 2006, Hong Kong had about 2.7 million Internet users (Hong Kong Industry Profiles, 2006).

Hong Kong also bears witness to the steady growth in the use of the Internet, mainly for communication and information purposes. In 2005, 15.5% of Hong Kong's companies had their own websites, up by 3.6% from 2004. These websites were mainly used to disseminate company information and collect customer feedback. Respectively, about 1.8% and 52.3% of Hong Kong's companies sold and received goods, services or information over the Net in 2005. The ACNielsen survey in the same year found that the favourite items or services of Hong Kong's online shoppers included books (35%), event tickets (23%), tours/hotel reservations (19%) and airline/tickets reservations (16%). The same survey indicated that most Hong Kong online shoppers used credit cards to settle their online purchases, placing Hong Kong as the second highest in the Asia Pacific and the third highest in the world in the use of credit cards for online purchases. In quantitative terms, the users of online business in Hong Kong increased from 96.5% in 2004 to 97.3% of the total business population in 2005 (Hong Kong Industry Profiles, 2006).

A number of B2B websites have been established in Hong Kong. They are usually created for standard commodity trading in resources such as steel, and for sourcing activities or supply chain management. A survey on e-Business adoption in Hong Kong, conducted by the Hong Kong Productivity Council in December in 2005, showed that the percentage of companies adopting higher levels of e-Business had increased to 24.2% from 20.5% in December, 2004. The survey generally indicated that more companies had moved up to higher levels of e-Business adoption. The government had responded by enacting the Electronic Transaction Ordinance legislation and by establishing the public infrastructure necessary to enhance public confidence in the legality and enforceability of e-transactions. Hong Kong Post and Tradelink are the country's recognised public certificate authorities for e-transactions. In order to promote the use of the Internet in Hong Kong, the government launched the Electronic Service Delivery Life Scheme (ESDLife), which includes Government-to-Citizen (G2C) and Government-to-Business (G2B) online services like the Electronic Tendering System (Hong Kong Industry Profiles, 2006).

During 2005/06, the Hong Kong government increased its IT expenditure to HK$4.47 billion, up from HK$3.75 billion in 2004/05. In April of 2005, Hong Kong reportedly had the best electronic business environment in Asia (Economist Intelligence Unit, 2005). In the same year, Hong Kong was ranked among the world's top 10 cities in digital governance by the Rutgers-SKKU. E-Governance Performance Index - an index of municipal websites worldwide - found that among Asian cities, Seoul (Korea) was the top-ranked city in terms of e-governance performance. The study was conducted jointly by the E-governance Institute of Rutgers University (Newark) and the Global e-Policy e-Government Institute of the Graduate School of Governance, Sungkyunkwan University, Korea. The Rutgers-SKKU E-Governance Performance Index is a systematic effort to evaluate digital governance in municipalities throughout the world. Based on the evaluation of 100 cities, the top 10 cities in 2005 in terms of percentage and in descending order were as follows (Rutgers-SKKU, 2006):

1. Seoul (81.70%)
2. New York (72.71%)
3. Shanghai (63.93%)
4. Hong Kong (61.51%)
5. Sydney (60.82%)
6. Singapore (60.22%)
7. Tokyo (59.24%)
8. Zurich (55.99%)
9. Toronto (55.10%)
10. Riga (53.95%).

The 2005 survey examined 100 municipalities from around the world, selecting the largest city in each of 98 countries with the highest percentages of Internet users. The Rutgers-SKKU E-Governance Performance Index uses 98 measures divided into five core areas: 1. Security and Privacy; 2. Usability; 3. Content; 4. Services; and 5. Citizen Participation. An overall score for each municipality (on a 100-point scale) is derived by providing equal weight to each of the five categories (Rutgers-SKKU, 2006).

With excellent telecommunications infrastructure, proximity to the Chinese mainland, and bilingual language capability, Hong Kong is able to act as the main gateway to e-commerce in the region, supporting the routing of Internet traffic and the flow of e-commerce. Many voice and data transmission services via the Internet, such as Internet Protocol over the telephone, have been developed in the country. Furthermore, the development and hosting of portals in Hong Kong have gradually transformed Hong Kong into an e-commerce hub for foreign companies trading in the region. Further efforts to promote e-commerce in Hong Kong are outlined in the government's Digital 21 Strategy consultation paper. The paper proposes five areas that require further action, namely facilitating the digital economy; promoting advanced technology and innovation; further developing Hong Kong as a focal point for technological cooperation and trade; enhancing the next generation of public services; and building an inclusive, knowledge-based society (Hong Kong Industry Profiles, 2006).

Elsewhere in Asia, India is gradually emerging as a powerhouse in e-commerce. One of its private companies, namely ITC, is a leader in e-commerce trade in the region, with annual revenues of approximately US$ 2 billion. The ITC International Business Division was created in 1990 as an agricultural trading company, and now generates US$ 150 million in revenues annually. The company initiated a project called "E-Choupal" ('Choupal' means gathering place in Hindi), which places computers with Internet access in rural farming villages. The computers, typically housed in farmers' houses, are linked to the Internet via phone lines or, increasingly, by a VSAT connection. Each computer serves an average of 600 farmers in 10 surrounding villages within about a five kilometre radius. Each E-Choupal costs between US$ 3 000 and US$ 6 000 to set up, and about US$ 100 per year to maintain. Using the system costs farmers nothing; however, the host farmer, called a Sanchalak, does incur some operating costs and is obliged to serve the entire community. In turn, the Sanchalak benefits from increased prestige and a commission paid for all E-Choupal transactions (OECD, 2004).

The farmers can use the computer to obtain and assess daily closing prices on local markets, track global price trends, or find information about new farming techniques. Accurate market price knowledge and market trends help them decide when and at what price to sell. They can also use the E-Choupal to order seed, fertilizers and other products (e.g. consumer goods) from ITC or its partners at lower prices than those available from village traders. The Sanchalak typically aggregates the village demand for these products and transmits the order to an ITC representative. Farmers benefit from more accurate weighing, faster processing time, prompt payment, and access to a wide range of information. By mid-2003, E-Choupal services reached more than 1 million farmers in nearly 11,000 villages. The E-Choupal serves as both a social gathering place for the exchange of information and as an e-commerce hub (OECD, 2004).

In 1940, a magnet manufacturing company by the name of Ni-roku was established in Japan. Through the efforts of an employee, a company website was launched in September, 1997 (www.26magnet.co.jp). By 2000, annual e-commerce sales of magnet products via the site had already generated more than US$ 700 000. The story goes that in 1996, a company employee bought a personal computer at his own expense to use the Internet. He then convinced the manager to launch the company's website. A small ICT service firm was consulted to help the company create the web pages, and the employee took training to enable him to update the site's content. In order to increase the number of visits to the site, the company used an online campaign and offline advertising. A free magnet offer campaign on the website combined with an online questionnaire to potential customers lured 1 000 visitors to the site during the first month. In response to requests from businesses as well as individual buyers, an online product catalogue was later added to the site (OECD, 2004).

AFRICAN DIGITAL ECONOMY INITIATIVES

Africa, as revealed in the e-readiness assessments already discussed, is still largely putting institutional mechanisms in place to take part in the global digital economy. There are, however, some exceptions. In Kenya, for example, millions of mobile subscribers without bank accounts can now make simple financial transactions using their phones. This was made possible through the launch of M-Pesa, a low-cost money transfer system from Safaricom (Kenya's main mobile service provider) in March of 2007. All that this transaction requires is a new generation SIM card, available for free from the mobile operator. Users are able to send money to all mobile phones, including those connected to rival operators, and receive money from other Safaricom subscribers. M-Pesa, as a form of microfinance banking, has the potential to at least partly replace cash-replacement services. It has been rolled out in Kenya in partnership with Citibank and the Commercial Bank of Africa. The service provides an affordable, fast and safe way to transfer money via SMS. This service is very important, especially to SMEs in rural areas. Subscribers need not have contact with the banks involved, and the banks likewise do not need the subscribers' details. M-Pesa is a valuable case study of digital money that entails the replacement of cash with electronic money. Its benefits lie in the fact that it is developed for the mass market, reduces transaction costs, and provides new functionality to the mobile phone (Rice, 2007).

M-Pesa (meaning mobile money) is expected to revolutionise banking in a country where more than 80% of the people are excluded from the formal financial sector. In time, M-Pesa will allow people to borrow and repay money, and make purchases. Companies will be able to pay salaries directly into workers' phones - something that has already attracted the interest of larger employers (such as Kenya's tea companies) whose workers often have to be paid in cash because they do not have bank accounts. Apart from transferring cash - a service in high demand among urban Kenyans supporting relatives in rural areas - customers of the Safaricom network will be able to keep up to 50,000 shillings (£370) in a "virtual account" on their handsets. With regard to security issues, the mobile provider assures users that in the event that someone's phone is stolen, the PIN system prevents unauthorised withdrawals. The main danger lies in sending cash to the wrong mobile number, which the recipient could redeem straight away (Rice, 2007). The M-pesa project is being watched closely by mobile operators around the world as a potential gateway into the multibillion pound international cash transfer industry, long dominated by companies such as Western Union and Moneygram. Remittances sent from nearly 200 million migrant workers to developing countries totalled £102bn in 2006, according to the World Bank. A similar initiative to M-Pesa reportedly exists in the Philippines, where two companies, Globe Telecom and Smart Communications, have been operating money transfers in the country since 2005.

LEARNING ORGANISATIONS' BEST PRACTICES IN THE DIGITAL ECONOMY

Researchers and academics define and use the term 'learning organisation' differently in literature. A learning organisation refers to an organisation in which people at all levels continuously increase their capacity to produce the results they need (Sutherland, 2003). A learning organisation can also be perceived as an organisation in which people can expand their capacity to create the results they wish, expand their thinking, and continuously learn, whether individually or collectively (Senge, 1993:3).

Given the implicit and explicit aspect of knowledge management (KM) in the definitions of a learning organisation, knowledge capital in the competitive operations of organisations becomes imperative. A learning organisation applies the principles of KM in harnessing its human capital. Consequently, a learning organisation is a knowledge intensive organisation. The World Bank Group (2008) recognises that highly skilled and flexible human capital is needed to compete effectively in today's dynamic global markets. In this respect, the ability to produce and use knowledge has become a major factor in development and is critical to a nation's competitive advantage. It is expected that the surging demand for secondary education in many parts of the world will create an invaluable opportunity to develop a workforce that is well-trained and capable of instituting knowledge-driven economic growth. The World Bank wishes to help countries adapt or transform their entire education systems to meet the new challenges of the "learning economy" through two complementaries, namely the formation of a strong human capital base, and the construction of an effective national innovation system [as discussed below] (The World Bank Group, 2008):

a) Formation of a Strong Human Capital Base

The World Bank recognises that part of the framework for knowledge-driven growth requires education systems to: impart higher-levels of skills on a greater portion of a nation's potential workforce, foster lifelong learning, and promote the international accreditation of the country's educational institutions. Efforts are therefore needed to provide quality and relevant education to a larger share of each new generation of young people through expanded secondary and tertiary education. It is also necessary to train/ retrain the existing labour force to foster new opportunities for those who were unable to complete secondary or enter tertiary education.

b) Construction of an Effective National Innovation System

The World Bank states that a national innovation system consists of a network of firms, research centres, universities, and think tanks that work together to take advantage of the growing stock of global knowledge; assimilate and adapt it to local needs; and create new technology when it is needed. The Ministry of Economic Development (2008) in Canada notes that the implication of the knowledge economy is that there is no way to leapfrog learning and knowledge-creation on the road to prosperity; hence these should be most nations' top priorities. A country's capacity to take advantage of the knowledge economy depends on how quickly it can become a "learning economy". Learning in this context not only implies using new technologies to access global knowledge, but also using them to communicate with other people about innovation. In such a learning economy, individuals, firms, and countries will be able to create wealth in direct proportion to their capacity to learn and share information and innovation. Through organisational learning, institutions and firms alike acquire tacit knowledge and experience. Such knowledge is not likely to be available in codified form; nor can it be acquired through formal education and training. What it does require, is a continuous cycle of discovery, dissemination, and the emergence of shared understandings

Through learning organisations, intellectual capital as a firm's source of competitive advantage can be harnessed. Intellectual capital is left over after suppliers, employees, creditors or shareholders and the government have been paid, and obsolete assets replaced. This implies that in order for companies to become learning organisations, they must be knowledge driven. They must also learn how to translate changes in intellectual capital into the worth of their businesses, and ultimately be able to see this in their balance sheets. A firm's intellectual capital is reflected in the ability of employees to continuously improve production processes as a source of competitive advantage (OECD, 1999).

CHARACTERISTICS OF A LEARNING ORGANISATION

Several authors have attempted to characterise a learning organisation, and when put together, have come up with an organisation that promotes a culture of learning, supports life long learning processes, and creates a conducive environment with supportive organizational practices; in which all the employees exchange information and ideas freely, and learning and creativity are nurtured and rewarded; and where openness, learning from mistakes, trust and creativity are the tenets of the organisation (Skyrme, 2003; Michael & Higgins, 2002; Brandt, 2003). These authors also underline the presence of key management processes that define the learning organisation. These management processes include strategic planning; participatory management; employee empowerment; competitor analysis; performance measures; and a reward and recognition system. Tools and techniques also characterise learning organisations. These tools and techniques are learning and creativity skills that support individual and group learning, problem solving, interviewing, brainstorming, organising information, and implanting new knowledge into mental models, among others (Skyrme, 2003; Brandt, 2003; Senge, 1993).

Learning organisations are also characterised by whether they use a systems approach or a more holistic approach to problem solving. The systems approach integrates disciplines and compares them to reveal their relationships. It perceives the organization as a whole, and the impact of any action is seen to reverberate through all parts of the system (Skyrme, 2003). Personal mastery as a characteristic of the learning organisation refers to self-assessment and personal learning, and is premised on the fact that organizational learning is dependent on its staff members' level of learning. Personal mastery has two components, namely that one must define what one is trying to achieve (a goal), and one must also have a true measure of how close one is to their goal (Senge, 1990:139). As a consequence, those who wish to achieve a high level of personal mastery will always continue learning. Learning organisations are also characterised by mental models - the ability to view reality relative to personal vision or perceptions, and reconciling both into a coherent understanding. Team learning also characterises a learning organisation. It proffers personal and career benefits because each member of the team draws talent, experience and knowledge from the other members of the team. All team members are expected to work together to achieve common goals. In the process, teamwork provides greater insight into individual differences and enables individuals to learn to work together cohesively (Larsen, 1996).

The operations within a knowledge or digital economy are mostly of a knowledge intensive nature. Part of the business of learning organisations is knowledge creation. Consequently, from the perspective of information and knowledge management, it behoves organisations to increasingly apply information and management to enhance their competitive value. With the digital economy expanding as the use of Internet increases in the business environment, it becomes even clearer why effective knowledge management in learning organizations should take centre-stage. In this environment, KM is also important as businesses strive to be more competitive, continue to downsize, and try to determine how to capture the knowledge of their staff (particularly that which cannot be codified). The success of most KM initiatives in organisations depends on whether or not they are learning organisations (Scarbrough et al., 1998).

CONCLUSION

A number of countries worldwide and some international agencies are taking initiatives to enable companies operating in their jurisdictions and beyond to transact business online and effectively take part in the digital economy. One such initiative involves identifying and deconstructing best practices so that they can be emulated by those countries or firms wishing to competitively participate in the global digital economy. For example, the International Chamber of Commerce provides a model of global trust in e-commerce and rules of trade and settlement. European and North American countries have achieved high levels of e-readiness to enable businesses in their jurisdictions to operate in the networked world. In South Korea, the government supports B2B and IT use in SMEs, while the Mongolian government is driving the e-commerce programme from the Prime Minister's office. New Zealand's government has developed tax laws on e-commerce, and Hong Kong's government does not tax profits from e-commerce. The European Union has shaped its regulatory environment to cater for e-commerce. In Denmark, education efforts in IT and e-commerce are aggressively marketed by government. Moreover, developed countries, especially in the OECD region, have embraced the concept of learning organisations to prepare their organisations and adapt them to the changing global digital environment. While concerted efforts are being undertaken to enhance e-business environments, there are several policy and managerial challenges and issues that first have to be confronted and addressed on the road to achieving a truly global digital economy. For this reason, the next chapter deals with challenges of the digital economy and how they can be ameliorated.

REFERENCES

Andersen, K. V., N. Bjørn-Andersen, and J. Dedrick (2003). Governance initiatives creating a demand-driven e-commerce approach: The case of Denmark. The Information Society, 19(1), 95–105. doi:10.1080/01972240309475

Bojic, O. (2001). *Mapping SMEs policies in the digital economy: Germany, Greece, and the UK: A guide to readers*. London: Information Society Technologies.

Brandt, R. (2003). Is this school a learning organization? 10 ways to tell. *Journal of Staff Development, 24*(1). Retrieved on April 10, 2008, from http://www.nsdc.org/library/publications/jsd/brandt241.cfm

Economist Intelligence Unit & IBM Corporation. (2004). *Scandinavia consolidates lead in fifth annual economist intelligence unit e-readiness rankings*. Retrieved on April 14, 2007, from http://www.store.eiu.com/index.asp?layout=pr_storyandpress_id=1260000726

EUROPA-Europe's Information Society Thematic Porta. (2007). A dynamic e-business environment. Retrieved on October 13, 2008, from http://ec.europa.eu/information_society/index_en.htm

European Commission. (2002). *eEurope 2002-accelerating EC*. Brussels: European Commission. Retrieved on December 19, 2002, from http://europa.eu.int/information_society/eeurope/action_plan/ecommerce/

Fan, Q. (2000). *On the way to e-commerce: Chinese experiences (1998-1999)*. Retrieved on February 23, 2007, from http://ausweb.scu.edu.au/aw2k/papers/fan/paper.html

Find. (2004). *E-Business in Taiwan*. Retrieved on October 13, 2008, from http://www.find.org.tw/eng/news.asp?msgid=120&subjectid=6&pos=0#top

Government of Canada. (2006). *Government online*. Retrieved on February 13, 2007, from http://www.ged-gol.gc.ca/rpt2006/rpt/rpt10_e.asp

Hong Kong Industry Profiles. (2006). *Buy and sell quality products from Hong Kong, the Chinese mainland, and Taiwan*. Retrieved on October 24, 2008, from http://sourcing.tdctrade.com ICTA of Mongolia. (2006). *E-Mongolia national programme: Sukbaatar-1 information and communication technology authority (ICTA)*. Retrieved on June 12, 2007, from http://www.comminit.com/en/node/148327

International Chamber of Commerce. The. (1997). *The ICC electronic commerce project (ECP)*. Retrieved on June 17, 2008, from http://www.iccwbo.org/home/electronic_commerce/electronic_commerce_project.asp

Jutla, D., J. Dhaliwal, and P. Bodorik (2002). Supporting the e-business readiness of small and medium-sized enterprises: Approaches and metrics. Internet Research, 12(2), 139–164.doi:doi:10.1108/10662240210422512

Kim, E. (2007). *E-Business support program in South Korea.* Seoul: E-commerce Assistance Division, Industrial Policy Bureau, Korea's Ministry of Commerce, Industry and Energy.

Larsen, K. (1996*). Learning organizations*. Retrieved on May 5, 2008, from http://home.nycap.rr.com/klarsen/learnorg/

Lefebvre, E., & Lefebvre, L. A. (2000). *SMEs, exports, and job creation: A firm level analysis*. Montreal: CIRANO and Polytechnique de Montreal.

Mann, C. L., Eckert, S. E., & Knight, S. C. (2000). *Global electronic commerce: A policy primer*. Washington, D.C.: Institute for International Economics.

Michael, T. S. C., and S. E. Higgins (2002). Library as a learning organisation. Libri, 52, 169–182. doi:10.1515/LIBR.2002.169

Ministry of Economic Development–Canada. (2008). What is the knowledge economy? Retrieved on July 22, 2008, from http://www.med.govt.nz/templates/multipagedocumentpage____17263.aspx

OECD. (1999). *The economic and social impact of electronic commerce--preliminary findings and research agenda*. Paris: OECD.

OECD. (2004). *ICT, e-business, and SMES.* Paris: Publications Service.

Rice, X. (2007). Kenya sets world first with money transfers by mobile. Retrieved on November 20, 2008, from http://www.guardian.co.uk/money/2007/mar/20/kenya.mobilephones

Rutgers-SKKU. (2006). *Rutgers-SKKU e-governance performance index ranks Seoul #1*. Retrieved on November 20, 2008, from http://i-policy.typepad.com/informationpolicy/2006/06/rutgersskku_ego.html

Scarbrough, H., Swan, J., & Preston, J. (1998). *Knowledge management: A literature review.* London: Institute of Personnel and Development.

Senge, P. M. (1990). *The fifth discipline.* New York: Doubleday.

Senge, P. M. (1993). *The fifth discipline: The art and practice of the learning organization*. New York: Random House.

Skyrme, D. (2003). *The learning organization*. Retrieved on May 31, 2008, from http://www.skyrme.com/insights/3lrnorg.htm

Sutherland, S. (2003). *The public library as a learning organisation*. Retrieved on May 31, 2008, from http://www.ifla.org/IV/ifla69/papers/111e-Sutherland.pdf

The New Economy Task Force. (1999). *Rules of the road governing principles for the new economy, The new economy task force report.* Retrieved on July 18, 2008, from http://www.ppionline.org/ppi_ci.cfm?contentid=1268&knlgAreaID=128&subsecid=174

The World Bank Group. (2008). *Education for the knowledge economy*. Retrieved on July 22, 2008, from http://web.worldbank.org/wbsite/external/topics/exteducation/0,contentmdk:20161496~menupk:540092~pagepk:148956~pipk:216618~thesitepk:282386,00.html

Washisu, E. (2003). Technological innovation and state of e-commerce use in Finland. *ECOM Newsletter, 31*. Retrieved on February 23, 2007, from http://www.ecom.jp/ecom_e/latest/newsletter_no31.htm

Section 8
Challenges of the Digital Economy, SMEs, and E-Readiness

This section consists of one chapter: Chapter 18 (Challenges of the Digital Economy). The chapter collates the challenges discussed thus far in each chapter of the book. The challenges are categorised by type as follows: Internet-related challenges, e-commerce challenges, political and jurisdictional challenges, policy, regulatory, and legal framework challenges, and challenges peculiar to SMEs (skills shortages, lack of awareness, online challenges, competition, consumer protection, taxation, trade investment barriers, etc.). The chapter aims to showcase the challenges SMEs are likely to face and offer suggestions on how these may be ameliorated.

Chapter 18
Challenges of the Digital Economy

INTRODUCTION

There are various challenges faced by SMEs in their endeavour to make more active use of the Internet and e-business. These challenges vary widely across different sectors of the economy as well as from country to country. The most commonly cited problems relate to being unable to apply the Internet to business; preferences for established business models; lack of an enabling environment (lack of ICT skills, poor network infrastructure); high costs associated with ICT equipment, networks, software, ongoing support, etc; and security and trust issues (the poor security and reliability of e-commerce systems, uncertainty of payment methods, etc) [OECD, 2004].

DOI: 10.4018/978-1-60566-420-0.ch018

INTERNET-RELATED CHALLENGES

SMEs participation in the digital economy is hampered by many infrastructure challenges, particularly those relating to the Internet. While big organizations can generally muster the resources they need for infrastructure development to participate in the Net economy, the challenge is to get SMEs on board by working around their resource and skills shortages (Ramayah et al., 2005; Payne, n.d.). Recently, a common resource-related problem has been SMEs' limited capacity to buy and sustain broadband. Information security also causes great concern to SMEs because they largely rely on vendors for advice and assistance; most do not have the capacity to implement their own security measures. This and other issues relating to security and privacy over the Internet is the most overwhelming barrier to the adoption of e-commerce (Norazah, 2001). Accepting change, brought about by adopting new technologies, also poses significant challenges to SMEs. Essentially, resistance to change is the fear and unfamiliarity that go with fundamental and radical changes to business processes (Fink, 1998).

Though the digital economy is largely predicated on the Internet, there is no central coordinating body for the Internet, and consequently it is difficult to know whom to address if there are concerns or when one wishes to complain. There are also increasing reports of cyber crime and other problems associated with online information security and privacy. The nature of the Internet and Web technology presents difficulties of its own, because the Internet is not a single behemoth, but rather a combination of related technologies that require multiple design decisions to meet users' needs. As such, the Internet requires servers, communication links, software, end user devices and content to transmit data, which current model-based user interface techniques cannot effectively provide (Grigorovici et al., 2003). The revolution of different Internet and Web-related technologies, particularly recent developments and improvements in distributed systems, portable and ubiquitous computing, and the general convergence of various computing and telecommunications technologies; have made additional demands on the quality of the user interface and its ability to enhance online interaction. There are also issues concerning the governance of the Internet itself. For example, the increase in electronic commerce is expanding the number of Internet addresses required, and this has consequently accelerated the need for further reforms of the Domain Name System (Coppel, 2000).

Businesses need assurance about using the Internet in transmitting confidential information in their day-to-day communications with suppliers and customers. Consumers similarly wish to be assured about the identity of the party at the other end of the transaction, and to know that any agreement made electronically is legally binding (The New Economy Task Force Report, 1999). Some companies also raise concerns about the Internet's performance, arguing that reliability cannot be guaranteed. Such companies make comparisons with Electronic Data Interchange (EDI) transactions over private value-added networks, which they say are better for transmitting important information because it is more likely to arrive at its destination on schedule and intact. They point out that with EDI, if or when problems arise, a single network service provider would be accountable and responsible for their resolution. In contrast, they argue that the Internet offers no such guarantees because there is no single entity responsible for ensuring that a message leaves one point and arrives, intact, at another. Businesses and consumers alike are wary of giving away credit card information, even over the telephone, because it may be stolen or misused. There is also uncertainty about the Internet's performance, reliability, and security, especially the security of transactions and the privacy of electronic information.

E-COMMERCE CHALLENGES

Challenges facing SMEs with regard to using e-commerce systems could be minimized if systems designers involved them in the design and development of user interfaces (Spollen, 2007). Getting an interface wrong is costly because private sector organizations lose money if their interfaces do not fulfill user requirements. Likewise, if e-government does not provide acceptable interfaces to its users, it will lay waste to taxpayers' money. The e-commerce model for transacting business in government and in the private sector is still largely immature and not yet widely understood. E-commerce implementation in a digital economy across jurisdictions poses great challenges, especially with regard to the complexities involving different legal and policy requirements. For example, e-business transactions across borders may require regional or international protocols for dispute resolutions. The virtual environments of electronic markets make it more difficult to determine who the contracting parties are, where an electronic commerce operator is based, and whether that operator is complying with all the relevant legal obligations and regulatory regimes. This can create (legal and regulatory) uncertainty about which jurisdiction is responsible or which law is applicable in disputed cases (Coppel, 2000).

As pointed out already, consumers in the digital environment have concerns relating to privacy, consumer protection, the security of credit card purchases, and order fulfilment and delivery. The absence of commercial codes and legal recognition covering areas such as the acceptance of electronic signatures and documents, contract enforcement, and greater certainty and liability for damages that may arise out of electronic transactions, limit the take-up of e-commerce, particularly in the B2B sphere. If a vendor is located in one jurisdiction and the consumer in another, it is difficult to determine which law should govern the contract. In e-commerce, customers interact with invisible providers, delivery processes, and process controls. This, coupled with the intangible nature of e-commerce, the complex legal structure required for security enforcement, the diffuse nature of the interactions, and the need to gather and secure customer information, create a lot of uncertainty, and mean that organisations have to significantly invest in developing and sustaining customer trust in the new economy (Coppel, 2000).

POLITICAL AND JURISDICTIONAL CHALLENGES

Transacting business electronically across different jurisdictions is complicated by the different interpretations of international trade laws/regulations, even when they are uniform and consistent. For instance, in the European Union and the United States, difficulties arise when the laws passed are interpreted and executed quite differently in each member country or state. The United Nations Commission on International Trade Law (UNCITRAL) is mandated to harmonize and unify the law of international trade. Businesses selling in the global marketplace must be aware of UNCITRAL and other legal entities for reasons ranging from taxation, to privacy and avoiding criminal proceedings. Businesses must also be aware of whether or not their business content is considered legal within the jurisdiction in which the business is operating, or which countries' legal processes are applicable in the target market (Jutla et al., 2002). The OECD has defined policies that member states have agreed should be used to govern taxation, namely neutrality, efficiency, certainty and simplicity, effectiveness, fairness, and flexibility. Taxation rules within e-business deal with two specific issues, namely a government's right to tax, and double taxation. The government's right to tax states that governments are allowed to tax business activities that happen within their state. Difficulties quickly arise because of the stateless and borderless nature of the e-business environment. The issue of double taxation is raised when the state finds that a business or individual is doing business in its state, and taxes them in addition to the state to which the business or individual belongs (Jutla et al., 2002).

E-commerce, especially for digital products, blurs the notion of geographical or tangible boundaries such as the location of supply or residence of the business. Because trade policies, such as the tax policy, are based on such distinctions, governments may find it difficult to determine jurisdictions and tariff revenue rights. The laws and regulations a consumer relies on for protection at home may not apply in the merchant's country. Indeed, in some quarters there are concerns that the ability of the Internet to transcend national boundaries could emasculate the ability of regulatory bodies to fulfil their objectives. There is therefore a need to update regulatory frameworks and strengthen co-operation between regulatory bodies to achieve the goals of economic regulations without jeopardising the efficiencies likely to be associated with the growth of e-commerce. Furthermore, with the proliferation of international e-commerce, there is uncertainty about the application of existing customs duties. B2C e-commerce shoppers are rarely informed about duties they are liable to pay, and vendors find it difficult to provide information on the wide range of customs regulations across countries. In such circumstances, the consumer is uncertain of the final cost and could encounter delivery delays as goods are held until customs clearance (OECD, 1999a, 2005a, 2005b 2005c). Moreover, consumption taxes are levied on the principle of taxation at the place of consumption and in line with the rates set in individual countries or states. E-commerce, however, has the potential to undermine the application of domestic and national tax rules, especially for products that can be digitised and delivered online.

POLICY AND LEGAL REGULATORY FRAMEWORK CHALLENGES

Access to broadband telecommunications is critical in regions that intend to grow or attract a wide variety of businesses. However, because the demand for broadband tends to be more heavily concentrated in larger and mid-sized metropolitan areas, telecommunications companies focus most of their initial investment in these areas. In most rural areas, lack of demand coupled with higher costs mean that companies often cannot make an adequate return on their investment. For the full realisation of the benefits of the digital economy, the following actions are necessary: privatisation and market access in the telecommunications sector; provision of an enabling policy and regulatory framework; standardisation of technologies and applications; ensuring universal access for all in society; ensuring open access to networks to facilitate a wider availability of products and services; improving the interoperability of systems; creating open and competitive markets in a deregulated environment; securing technology systems - ensuring privacy and authentication of business and personal transactions; and protecting the privacy of personal information and intellectual property (The New Economy Task Force Report, 1999).

Companies operating in the digital environment fear that governments may over-regulate or censor the Internet by limiting content or complicating the ways in which buyers and sellers conduct business electronically. The US government, which has made great strides in the Net economy, supports the development of the Internet as a market-driven, unregulated platform. This means that governments should refrain from passing regulations that govern the Internet. The US government believes that where applicable, rules for Internet behaviour should be set through private, collective action rather than through government regulation. The aim of these rules should be to empower consumers to protect their own privacy, control the content they see, and protect themselves against inappropriate commercial behaviour. Competition and consumer choice should be the guiding principles of Internet commerce (The New Economy Task Force Report, 1999).

The role of government in a digital economy should be to provide an enabling environment that focuses on removing constraints to ensure the survival and growth of businesses that invest in it. As a result, governments need to create progressive economic policy frameworks that encourage the digital economy's growth and development. A policy framework for the digital economy should empower people with information that gives them control over their lives (The New Economy Task Force, 2000). Moreover, national policies to accelerate e-commerce & e-business applications should focus on: the promotion of B2B applications; promotion of e-marketplaces and helping SMEs participate in these markets; promotion of SME-support networks aimed at e-business application; raising SMEs' awareness of the benefits of ICT and Internet usage; organising training activities in order to eliminate ICT skills shortages and develop digital literacy; establishing Internet shops to provide e-commerce and e-business portal services; and ensuring that governmental bodies play their role in developing and extending e-commerce and e-business tools (Cabbar, 2002). Governments should also focus on creating equal opportunities by: ensuring that each individual and enterprise has access to information and information networks; establishing secure and confidential networks for e-commerce and e-business applications; ensuring that all the legislations relating to e-commerce and e-business are revised or renewed; establishing infrastructure and related services for e-commerce and e-business applications; measuring the effects of e-commerce & e-business on the entire economy; and developing positive aspects for the growth of the economy.

The business environment should encourage businesses to be transparent in terms of who they are, what they are doing and how they will serve their customers, because customers are not happy with their e-commerce experience when they lack adequate information (International Business Law Services, 2005). Trustmarks or Seals of approval are currently one of the chief ways of promoting consumer confidence and self-regulation in e-commerce. Usually, independent organizations (known as Code Owners) establish standards, known as Codes of Practice, for conducting e-commerce and certifying that particular online businesses, referred to as

Code Subscribers, have met those standards. A Code Subscriber is then permitted to display the Code Owner's Seal or Trustmark on their website. In this framework, Alternative Dispute Resolutions exist for out-of-court dispute settlement (International Business Law Services, 2005). Most Trustmark programs usually cover: the identity of the parties involved in transactions, server host identity, products and services, qualitative characteristics, warranties, contractual conditions, supply restrictions, delivery conditions, order error protections, payments, customer service, claims procedures, dispute resolution, advertising, reference legislations, consumer protection, security, website protection, payment data protection, and personal data protection.

Governments shoulder great responsibilities in developing an environment that is conducive for e-business to thrive. For one, they need to facilitate business models that enhance trustworthiness. They also need to enhance infrastructure deployment; implement secure networks; proffer adequate funding; provide tax relief in the purchase of PCs; standardise e-commerce systems across government and between government and the private sector; use alternative connectivity solutions, such as mobile technology; and promote the rigorous exchange of information and policy coordination among different nations (IDRC, 2007; Ministry of International Trade and Industry-Japan, 1997).

Governments must also take prudent policy actions to promote the digital economy by providing workers and communities with the skills and tools they need to successfully participate in the digital economy; enabling all those not yet engaged in or benefiting from the new economy to have access to the tools and resources they need to successfully participate; providing every citizen with access to continuous and affordable life-long education; and investing more in knowledge infrastructure, such as world class education, training and life-long learning, and maintaining science and technology standards, among other intangible public goods (The New Economy Task Force Report, 1999).

In their efforts to create an enabling environment for the digital economy, governments should further:

- Avoid policies and regulations that would inhibit the growth of the Internet or slow progress by protecting the business interests of those threatened by the digitization of the economy
- Craft a legal and regulatory framework that supports the widespread growth of the Internet and high speed telecommunications in areas such as taxation, encryption, privacy, digital signatures, telecommunications regulation, and industry regulation
- Help spur on the future growth of the Internet by co-investing in information infrastructure, including the Next Generation Internet (NGI)

- Strive to reduce or eliminate tariffs and unnecessary price regulations
- Become as responsive and flexible as the economy and society with which they interact
- Develop public policies that properly account for the diverse natures and uses of knowledge.

Perhaps countries could learn from the U.S. National Science Foundation (2005), who are of the view that an expanding environment for creating and managing the digital economy requires attention to a wide range of policy issues, including public investment priorities, program design, dissemination of research results, and technology transfer.

CHALLENGES PECULIAR TO SMEs

Although SMEs are affected by challenges relating to infrastructure and of a generic nature in the digital economy marketplace, they also face unique challenges. The preamble to case studies of e-business, focusing primarily on SMEs, at the American Conference on Information Systems, held annually since 2001, observes that e-business applications are often still very expensive for SMEs (Ehrlich, 2005) Furthermore, a lower degree of management and information technology expertise and a lower number of transactions supported by the applications are other limitations facing small-sized enterprises Rapid changes in society, coupled with the concise processes necessary to create a successful environment, add to the complexity of the problems experienced by SMEs. SMEs are often unable to deal with the complexities of internal and external communication and aligning the business with technological advancements to ensure that they remain competitive (Ehrlich, 2005).

Compared to larger businesses, SMEs face more difficulties when trying to find the capital they need to invest in new technology. They also have problems diverting staff time to research and deciding what aspects of electronic commerce make more sense to them. SMEs are hampered by their lack of business skills and experience, and the poor quality of their business plans. They also lack collateral when seeking micro-credit from banks. The importance of micro-credit to assist small- sized enterprises is widely acknowledged. During the Micro Credit Summit at the Africa Advocacy Forum II (held from the 10th - 13th of November, 2002 at the UN headquarters in New York), Kofi Annan, the former UN Secretary General, suggested that it is imperative to look at the pivotal role that sustainable microfinance could play, in reaching the Millennium Development Goal of halving poverty by the year 2015 (WSIS, 2003).

Although ICTs have been recognized as important assets in providing competitive advantage in global business environments, SMEs still face various obstacles that hamper their ability to enhance their competitive status. Small- sized enterprises are resource poor, with limited access to external data to enhance their ICT awareness in general and ICT opportunities in particular. In India, SMEs are slow to adopt IT to improve the efficiency and quality of their products, with only 8-10% of the over three-million SMEs having ventured into IT usage (Rajalakshmi, 2004).

The absence of internal specialists and pressure on management's time within SMEs result in firms being considerably reliant on the knowledge and expertise of external IT suppliers and IT consultants, which is a costly and unsustainable practice (Chau, 1996:187-9; Raymond, 1989). Moreover, despite the falling prices of ICTs, considerable investment is still needed on the part of SMEs in terms of both time and money (Cragg and King, 1993:47-52). The SMEs' efforts to attain the e-readiness level they need to participate in the digital economy are further hampered by the costs associated with the transformation to a higher e-readiness status. The principal categories of quantitative and qualitative costs of e-readiness include investment costs, maintenance costs, a steep learning curve, and Internet security costs, among others (Agriculture and Food Canada, 2001). The Canadian E-business Opportunities Roundtable (2001), in an '*Analysis of barriers impeding e-business adoption among Canadian SMEs*' identified several barriers that SMEs face, including the lack of information and education; uncertainty with regard to the costs and benefits of adoption; issues of access; the lack of strategic resources; and concerns with privacy and security. Barriers were also identified with regard to the lack of management commitment, low return on investment, and the insufficient availability of suppliers, both in terms of strategic advice and technological development. Moreover, SMEs were found to face more difficulties than larger companies when it came to introducing and effectively using new technologies; accessing R&D funds arising from co-operation with research centres or universities; and obtaining formalised management tools, to name a few.

SMEs also suffer from inabilities to prioritise, inadequate awareness, lack of familiarity with online access, and a preference for maintaining current business models. A World Bank study on 632 Black micro-enterprises in 1990 in South Africa that aimed to establish the principal constraints facing small-sized enterprises, found that illiteracy was the main factor that prevented SMEs from relying on printed sources of information (Ntsala, 2000:173). In another study of small-sized enterprises in KwaZulu-Natal and the Northern Province of South Africa, it was found that a number of entrepreneurs could not achieve a sound financial base because they could not get financial assistance from most financial institutions at the time- their requirements effectively cut out most members of the SME sector

(Ntsala, 2000). Ntsala (2000) also found that the lack of training initiatives that focused on small-sized enterprises' development was a further drawback to SMEs' development in the two regions. A study on the e-readiness of SMEs in the Philippines revealed that there were common problems facing the firms, such as the under-utilisation of PCs and other ICT tools, high costs of connectivity, continued migration of IT professionals, and the lack of laws and policies covering IT in the country. Moreover, SMEs were found to suffer from high production costs and a lack of adequate information delivery systems, causing the failure of most enterprises (Sachs, 2003; Gouws, 1997).

A review of the SME sector in Turkey in 2003 with regard to the level of e-commerce revealed that most commercial banks and investors regarded SMEs to be high risk borrowers because of insufficient assets and low capitalisation, their vulnerability to market fluctuations, poor accounting records, and inadequate financial statements. Moreover, SMEs could not fully compete because they did not have access to trade-related information. SMEs also did not know where the markets were for their products, and could not make their products known to potential customers because they lacked or could not afford the necessary equipment and software to do so (Gungen, 2003). The review revealed that SMEs' competitiveness was also hampered by inefficient institutional infrastructure; a poor legal and regulatory environment; limited management skills; and the inability to identify, locate, evaluate and apply information.

A study undertaken in Botswana by Mutula (2005) found that SMEs had a very negligible proportion of market share with regard to their core business. Comparatively, most large enterprises claimed to enjoy a sizeable chunk of market share, ranging from 20-71%. The study found that SMEs' insignificant local market share made them less competitive in the global digital economy. SMEs were also found to face various other obstacles, such as the high cost of bandwidth; limited legislation dealing explicitly with the protection of personal privacy or personal data; limited capacity to safeguard the right to have access to information, particularly information that affects one's self or one's business; and weak legislation regarding access to government information. SMEs in the country also faced challenges relating to the poor quality of telecommunications networks; high cost of telephones; frequent Internet downtime; slow Internet access speeds; high taxation, making ICT expensive; high cost of software licensing; congestion in mobile phone connectivity, especially at the end of the month; frequent dropped connections on mobile phones; and the lack of adequate technical support.

De-Kare Silver (2000), in a general study on the electronic selling of services and products, identified the following obstacles to SMEs: inadequate manpower and financial resources; lack of technical expertise to manage online transactions;

lack of senior management support; and the lack of awareness of the benefits of e-commerce for business. A similar study by the Department of Enterprise, Trade and Employment (2004) in Ireland identified a number of obstacles to the greater and more effective use of ICTs by SMEs, including: the lack of appreciation on the part of owners/managers of the contribution of ICTs to their businesses; lack of appropriate internal IT resources; and the relatively high costs associated with investments in ICTs. SMEs are also constrained by their export capacity for various reasons, including the lack of incentives from government; inadequate marketing of their products; lack of awareness of other (regional and international) opportunities; lack of government support in exploring international business opportunities; the small-size of companies, often forcing them to operate within local niches; and stiff competition from larger enterprises and international businesses.

Lee et al. (1994), in an APEC survey of SMEs in Asia, observed that the three greatest obstacles preventing SMEs from achieving their export potential appeared to be the difficulties associated with obtaining finances to send goods and the general reluctance to enter into international trade markets. SMEs also face various trade impediments, such as varying standards; restrictive and sometimes punitive customs regulations; lack of intellectual property protection mechanisms; unfair government subsidies to competitors; difficulty in obtaining working capital; lack of available resources; limited overseas market intelligence; complexities that arise in meeting government regulatory requirements; shortage of funds for setting up business channels abroad; and the general lack of experience in international business.

Without strong education and support programmes, SMEs tend to avoid the 'unknowns' of doing business in foreign markets. Unlike large enterprises, many of which already operate in global markets through well established networks, the export potential of SMEs tends to be constrained (Moodley, 2001:90; Southwood, 2004). Most SMEs are limited in their capacity to partner with multinational companies or compete externally because their equity base may potentially shift over time. Schmidt (1998) emphasises the need to create awareness amongst SMEs about the potential benefits of partnerships, support SME participation in appropriate international trade fairs, and encourage foreign cooperation in order to enhance the export capacity of these enterprises.

Most SMEs tend to have CEOs with limited prior ICT experience. In turn, these CEOs are more inclined to employ lower skilled personnel because they are cheaper and the kinds of businesses they run do not warrant high skilled personnel. This practice has implications on the ability of the companies to compete effectively in both local and international markets. In Britain, SME growth potential is linked directly to the levels of education and experience of the founders and managers. There is a propensity for managers with higher qualifications to be more active in their management and staff development attitudes than those with lower qualifications (Gray and Lawless, 2000). British SMEs, without the buffers of their larger corporate cousins, rely on the collective skills of their workforce for business performance and for IT user skills - the fourth basic skill alongside communication, literacy and numeracy. The findings of an e-skills UK study on IT skills in the workplace in 2003 revealed very worrying discrepancies. For example, it was found that many employees lacked the IT skills they needed to effectively perform their roles, which had a direct effect on business productivity. Moreover, skills constraints influenced company performance indirectly by restricting the use of IT that could be beneficial to the SME. If, for example, users were not aware of viable IT shortcuts to business operations, such as customer ordering or document creation and presentation, it took them much longer to produce a professional body of work. Furthermore, poorly skilled IT users made up to 50% more errors than their well-trained peers (Price, 2003).

Another study in the UK found that awareness, attitudes, capabilities and investment were important factors in the take-up of ICT by SMEs. The study found that many SMEs, particularly micro-businesses, could not derive the maximum benefits from ICT because of their relative lack of knowledge and resources, including skills that they needed to become expert purchasers and appliers of new technology (Sector Skills Development Agency [UK], 2004). Dagdilelis and Satratzemi (2003), in a case study of a nationwide system for training SMEs for ICT innovation in Greece, found that the employees of SMEs did not, in many cases, have positive attitudes towards ICT. They noted that particular characteristics of small-sized enterprises and their staff had an effect on their approach towards ICTs, especially in terms of the knowledge they needed to deal with a variety of situations in the various enterprises. A study on the e-readiness of SMEs in the textile sector in Egypt found that SMEs did not understand the ways in which the Net could enable them to operate their businesses more efficiently and cost effectively. The study revealed a limited understanding of the complexity of e-commerce by SMEs. Moreover, their ICT skills were inadequate, and there was a lack of strategic planning for the long term development of their businesses (Rizk, 2004; Ramsey et al., 2003:253; Schmidt, 1998).

Often, SMEs are constrained by the poor quality of Internet connectivity. Moreover, there are usually minimal legislative structures to assure the entrepreneurs of the security of their online transactions. In Egypt, for example, Rizk (2004) reported that SMEs within the textile sector relied mostly on dial-up access to establish Internet connectivity. This was mainly because of the high costs associated with full Internet connectivity. The quality and speed of a network is also important in determining how the network is used. A large number of mainline faults, poor connections, dropped connections and packet loss could render any network useless or operationally sub-optimal, thus discouraging the use of and investment in new technologies. SMEs sometimes suffer from the general lack of a competition policy to enhance consumer protection, thus hampering business growth. A consumer protection policy is necessary to protect against un-competitive practices and promote fair play in the market. A protection regime ensures that programmes or policies are available to educate consumers about e-commerce in order to foster consumers' awareness of online activities. In the absence of a consumer policy, innovations may be copied for free, thus robbing the innovators of their future revenue. OECD countries have developed guidelines for consumer protection in the context of e-commerce, and these have formed the basis for government consumer protection programmes internationally (OECD, 1999a).

Small-sized enterprises often express concern, especially in developing countries, about the high taxation that is levied on ICT goods and services. The high taxes levied on these products affect the growth, profitability and competitiveness of their enterprises. There are usually two main costs associated with taxation, namely import duty and value added tax, the latter of which gets passed over to the consumer. If SMEs are to compete in the international market, their goods and services need to be competitive in terms of pricing and quality. OECD (2003), in a study on the impact of taxation and other regulations on small-sized enterprises in Niger and Swaziland, established that taxation is a great burden to SMEs, although entrepreneurs in the two countries also felt that the lack of demand and/or difficulties in obtaining finance or raw materials were often the main constraints they faced.

Mutula (2005) found that in Botswana, SMEs faced various obstacles to trade and investment. One such obstacle was the need for a trade policy that gives incentives to SMEs to enable them to compete effectively with bigger companies, and a simplified industrial licensing system that would speed up the issuance of licenses as well as bring the services closer to the businesses. Other obstacles mentioned include bureaucratic tendering procedures, the amount of time it takes to obtain licenses, high taxation, the lack of skilled labour, and the lack of investment incentives. Governments, especially in the developing world, should offer incentives to SMEs by establishing special zones where regulatory or pricing constraints are more flexible. UNIDO (2005) points out that while SMEs constitute more than 90% of all businesses in Africa, they suffer from a high mortality rate as they struggle against mountain-sized odds. It should be remembered that most SMEs operate at the lowest rung of the cash economy, competing and surviving on the basis of low costs, low prices and low quality.

SMEs often have difficulty in using electronic commerce because they have a poor understanding of e-commerce techniques and the technology required to effectively take part in it. Awareness-raising activities would help diffuse initial information about ICTs and e-commerce, and should therefore be a major policy focus. Governments could also provide, through their various portals, information covering a wide range of e-business issues with links to related support services, seminars and workshops that demonstrate the possible integration of ICTs and e-commerce into SMEs' processes. Case studies could also provide practical business pointers to managers and employees. Televised award ceremonies may also act as incentives by giving high visibility and publicity to the highest achievers in e-commerce. 's National Office of the Information Economy has developed an e-business guide, essentially a comprehensive online resource that is aimed at business people who want a quick and easy explanation of e-business - what it is, its benefits and what is needed to get started. Such a guide can help SMEs self-assess the costs, risks and opportunities associated with e-business. It can also help SMEs recognise the importance of strategic management and ICT skills development.

CONCLUSION

There are a number of challenges facing both small-sized and large enterprises and governments that must be addressed in order to enhance the digital economy environment. Most of these challenges largely stem from the digital divide, such as limited access to technology, inadequate skills, poor network connectivity, high cost of access, etc. One challenge in particular relates to online security and privacy, which is exacerbated by the fact that the Internet does not have an effective central coordination and control mechanism. With whom, for instance, does one lodge complaints if one's rights are violated online? Moreover, there is a general lack of awareness about the potential of technology and in particular, what e-commerce can offer, especially to SMEs. There is therefore a need to educate business enterprises on the potential of ICT in business transactions. Governments should be encouraged to work through bilateral and multilateral arrangements to address cross jurisdictional legal frameworks that enhance cross-border e-commerce. Furthermore, the peculiarities of SMEs demand that the problems they face should be given special treatment if they are to take part in the digital economy. The success of nations in the Net economy is predicated on the level of their SMEs' e-readiness. Policy and managerial interventions to stem the challenges facing SMEs are needed to enable them to participate in the global digital economy. Without the full participation of SMEs, this hyped-up economy may become the economy that never was.

REFERENCES

Agriculture and Food Canada. (2001). The e-readiness of Canadian Agriculture and Food Canada SMEs. Retrieved on March 21, 2004, from http://www.agr.gc.ca/misb/fb/food/ecommerce/eready/introduction_e.html

Cabbar, Y. (2002, May 2). Entrepreneurship and SME development. *Opening statement at Black Sea Economic Cooperation Workshop on Internet, E-commerce, and SMEs*, Ankara, Turkey.

Canadian E-business Opportunities Roundtable. (2001). *Analysis of barriers impeding e-business adoption among Canadian SMEs.*

Chau, P. Y. K. (1996). An empirical assessment of a modified technology acceptance model. *Management Information Systems, 13*(2), 185-205.Coppel, J. (2000). *E-commerce: Impacts and policy challenges. Economics department working papers no. 252*. Retrieved on July 20, 2007, from http://www.oecd.org/eco/eco

Cragg, P. B., and M. King (1993). Small-firm computing: Motivators and inhibitors. MIS Quarterly, 17(1), 47–60. doi:10.2307/249509

Dagdilelis, V., & Satratzemi, M. (2003). Implementing a nationwide system for training very small enterprises for ICT innovation: The Greek case. *Educational Technology & Society, 6(1)*. Retrieved on March 5, 2005, from http://ifets.ieee.org/periodical/vol_1_2003/dagdilelis.html

De-Kare Silver, M. (2000). *E-Shock*. London: Macmillan Press Ltd.

Department of Enterprise. Trade, and Employment (Ireland). (2004). *Ebusiness strategy*. Retrieved March 4, 2005, from http://www.entemp.ie/enterprise/support/ebusinessstrategy.htm

Ehrlich, D. M. (2005, August 11-14). A study concerning small manufacturing business within B2B e-commerce value chain. In *Proceedings of the Eleventh Americas Conference on Information Systems*, Omaha, NE.

Fink, D. (1998). Guidelines for the successful adoption of information technology in small and medium enterprises. International Journal of Information Management, 18(4), 243–253. doi:10.1016/S0268-4012(98)00013-9

Gouws, E. (1997). Entrepreneurship education: An education perspective. South African Journal of Education, 17(3), 143–149.

Gray, C., & Lawless, N. (2000). *Innovations in the distance development of SME management skills*. Retrieved March 4, 2005, from http://www.eurodl.org/materials/contrib/2000/gray.htm

Grigorovici, D. M., Schement, J. R., & Taylor, R. D. (2003). *Weighing the intangible: Towards a framework for information society indices.* University Park, PA: Institute for Information Policy, Pennsylvania State University.

Gungen, T. (2003). *E-Finance for SMEs in Turkey: Yapi Kredi Bank.* Retrieved on November 27, 2003, from http://r0.unctad.org/ecommerce/event_docs/xmefinoct2001/gungen.pdf

IDRC. (2007). *Problems encountered in e-commerce initiatives in Southern Africa: what lessons can we learn?* Retrieved on February 5, 2007, from http://network.idrc.ca/en/ev-11603-201-1-DO_TOPIC.html

International Business Law Services. (2005). *Electronic invoicing in Germany*. Retrieved on February 5, 2007, from http://www.ibls.com/document_top.asp?d=215

Jutla, D., J. Dhaliwal, and P. Bodorik (2002). Supporting the e-business readiness of small and medium-sized enterprises: Approaches and metrics. Internet Research, 12(2), 139–164.doi:doi:10.1108/10662240210422512

Lee, C. J., Chi-Keung, L., & Hwang, T. S. (1994). *The APEC survey on small and medium enterprises* (pp. 1-15). Singapore: APEC Secretariat.

Ministry of International Trade and Industry–Japan. (1997). *Towards the age of the digital economy--for rapid progress in the Japanese economy and world economic growth in the 21st century*. Tokyo: Ministry of International Trade and Industry Japan.

Moodley, S. (2001). Impact of electronic commerce and small exporting firms in the South African wooden furniture manufacturing sector. Journal of Information Technology Impact, 2(3), 89–104.

Mutula, S. M. (2005). *Assessment of the e-readiness of small and medium-sized enterprises in the ICT Sector in Botswana, with special reference to information access*. Unpublished doctoral dissertation, University of Johannesburg, Johannesburg.

Norazah, M. S. (2001). Malaysian Internet users' motivations and concerns on shopping online. Malaysian Journal of Library & Information Sciences, 6(2), 21–33.

Ntsala, M. (2000, October 26-27). Information delivery systems for small business. In *Proceedings of First Biannual DISSAnet Conference* (pp. 167-176). Pretoria: The Centre for Information Development, University of Pretoria.

OECD. (1999a). *The economic and social impact of electronic commerce--preliminary findings and research agenda*. Paris: OECD.

OECD. (1999b). Securit-e-business. *OECD Observer, 219*. Retrieved on August 7, 2008, from http://www.oecdobserver.org/news/fullstory.php/aid/175/Securit-E_business.html

OECD. (2003). *The impact of laws and regulations on micro and small enterprises in Niger and Swaziland*. Retrieved on March 5, 2005, from http://www.oecd.org/LongAbstract/0,2546,en_2649_37461_33779951_70666_119684_1_37461,00.html

OECD. (2005a). *Communications outlook*. Paris: OECD.

OECD. (2005b). *Telecommunications databases*. Paris: OECD.

OECD. (2005c). *ICT key indicators*. Retrieved on August 7, 2008, from www.oecd.org/sti/ICTindicators

Payne, J. E. *E-Commerce readiness for SMEs in developing countries: A guide for development professionals*. Washington, D.C.: Academy for Educational Development.

Price, K. (2003). *IT user skills in SMEs*. Retrieved on March 5, 2005, from http://www.trainingreference.co.uk/skills/it_user/it_user_skills.htm

Rajalakshmi, R. (2004). *SMEs lagging in IT e-readiness. Enterprise resource planning update*. New Delhi: Software Technology Parks of India (STPI).

Ramsey, E., P. Ibbotson, J. Bell, and B. Gray (2003). E-Opportunities of service sector SMEs: An Irish cross-border study. Journal of Small Business and Enterprise Development, 10(3), 250–264. doi:10.1108/14626000310489709

Raymond, L. (1989). Management information systems: Problems and opportunities. International Small Business Journal, 7(4), 44–53. doi:10.1177/026624268900700404

Rizk, N. (2004). *E-Readiness assessment of small and medium enterprises in Egypt: A micro study.* Cairo: American University. Retrieved on August 9, 2004, from http://www.sba.luc.edu/orgs/meea/volume6/Rizk.htm

Sachs, J. D. (2003). *Readiness for networked world.* Retrieved on November 28, 2003, from http://cyber.law.harvard.edu/readinessguide/intro.html

Schmidt, J. (1998). Enhancing the competitiveness of SMEs in transition economies and developing countries in the global economy and their partnership with SMEs of OECD countries. Retrieved on March 1, 2005, from http://www.conferenzabologna.ipi.it/Eng/dopo_la_conferenza/documenti/WK4_eng.pdf

Sector Skills Development Agency (UK). (2004). *Skills for business*. Retrieved on March 3, 2004, from http://www.ssda.org.uk/ssda/Default.aspx?page=740

Southwood, R. (2004). *Botswana special-ICT sector to enter global market by 2004*. Retrieved on December 6, 2002, from http://www.balancingact-africa.com/news/back/balancing-act_87.html

Spollen, M. (2007). *Challenges for e-government*. Retrieved on February 3, 2007, from http://infocentre.frontend.com/infocentre/articles/challengesforegovernment.html

The New Economy Task Force. (2000). *Making the new economy grow an action agenda*. Retrieved on July 21, 2008, http://www.ppionline.org/ppi_ci.cfm?contentid=1490&subsecid=123&knlgAreaD=107

The New Economy Task Force Report. (1999). *Rules of the road governing principles for the new economy, the new economy task force report*. Retrieved on July 18, 2008, from http://www.ppionline.org/ppi_ci.cfm?contentid=1268&knlgAreaID=128&subsecid=174

UNIDO. (2005). *SMEs in Africa survive against all odds*. Retrieved on March 5, 2005, from http://www.unido.org/en/doc/3926

U.S. National Science Foundation. Organisation for Economic Cooperation and Development, Research Directorate-General, European Commission, Information Society Directorate-General, European Commission, & U.S. Interagency Working Group on IT R&D University of Michigan. (2005). *Advancing knowledge and the knowledge economy*. Retrieved on July 21, 2008, from http://advancingknowledge.com

WSIS. (2003). Draft declaration of principles. Geneva: WSIS Civil Society Working Group. Retrieved on November 7, 2004, from http://www.wsis-pct-org/prepcom2-declaration.html

Section 9
Conclusion, Policy, and Managerial Recommendations

This is the final section of the book and consists of one chapter: Chapter 19 (Conclusion and the Way Forward). The chapter summarises the main issues discussed in the book and proffers policy and managerial recommendations that can empower SMEs to compete effectively in the global digital economy.

Chapter 19 Conclusion and the Way Forward

SUMMARY AND CONCLUSION

The primary focus of this book has been on digital economies, SMEs and e-readiness. With the exception of this chapter, which provides the book's summary and conclusion, the eighteen chapters that make up this book collectively cover conceptual aspects of the core subjects discussed, i.e. the digital economy's components and infrastructures; the digital divide and its implications to SMEs, e-readiness and the digital economy; e-readiness measurement tools and methods; e-commerce and e-business; e-government; content, knowledge and e-records management; e-readiness of SMEs in the digital economy; information needs and behaviour of SMEs in the digital economy; capacity building - SMEs and the digital economy; globalisation of the digital economy; trends and best practices in the digital economy; and challenges of the digital economy.

DOI: 10.4018/978-1-60566-420-0.ch019

The evolution of the Internet and the World Wide Web (WWW) is mostly having a positive impact on the growth of the digital economy. The Internet has made it possible to connect people to people, people to businesses and people to government, resulting in a networked world. The information explosion and the increasing recognition that information and knowledge are key ingredients in the modern economy have given further impetus to the development and growth of the digital economy. When well harnessed, the digital economy has the potential to enable business enterprises to penetrate and compete in the international market, particularly in the case of SMEs, which have in the past been excluded from mainstream global trade.

The digital economy environment consists of various components and infrastructures, including telecommunications and financial services, computing infrastructures (hardware, software, storage devices, end user technologies, and artificial intelligence systems), the Internet and WWW, websites, and much more. Emerging technologies such as Wi-Fi, VOIP, Bluetooth, Next Generation Internet, Web 2.0, etc; are also already defining the success or failure of modern enterprises in the digital economy. These technologies all rely on the adequate and continuous supply of electricity for their operations. The inadequacy of any of these components can quickly give rise to a digital divide between those enterprises that have them (especially larger enterprises in developed countries) and those that don't (e.g. SMEs in developing countries). The availability or otherwise of adequate infrastructure and other components determines the e-readiness or preparedness of the enterprises to participate in the global digital economy.

The concept 'e-readiness' is emerging as a very important measure of which business enterprises are prepared to take part in the digital economy. The concept originated in the late 90s to provide a way of measuring the breadth and depth of the digital divide between the developing and developed worlds. The concept generally refers to the preparedness of a community or country to participate in the networked world. An e-ready society or country would therefore have, for example, high speed access to the Internet; pervasive connectivity in schools, government offices, businesses, homes, hospitals, etc; capacity to ensure user privacy and online security; and an enabling policy and regulatory framework that allows digital signatures, official communication through e-mail, etc. Comparative measures of the e-readiness of countries worldwide are usually published annually, and generally show that countries in North America and Europe are leaders in e-readiness stature, and are far ahead of their counterparts in developing or transitional economies. Countries with high levels of e-readiness are found to be performing well in the digital economy.

The e-readiness of communities or countries is assessed using various tools and methodologies. However, most of the existing e-readiness assessment tools were designed to investigate e-readiness at macro-sectoral level and not for assessing micro-sectoral entities such as SMEs. The existing macro e-readiness tools and methods have varied definitions of the concept, with no standard definition of e-readiness and methods of assessment. The choice of the tool depends mainly on what it is to be used for and who is carrying out the investigation. The existing tools focus more on ICTs and the business environment, and pay little to no attention on information access, without which it is impossible to connect firms to the global networked world. The macro e-readiness tools also fail to address the impact of e-readiness on business organisations, and often do not provide suggestions on what actions to take to move a country up the e-readiness ladder. The tools are also largely quantitative, with their main method of data collection being through a short survey questionnaire, and as such, very little information is captured about people-centric issues such as perception, attitudes, skills, and awareness. Without the capture of detailed qualitative aspects of e-readiness, a lacuna exists in decision making processes on how improvements can be made to enhance the status of business organisations. The other shortcoming of the existing e-readiness tools and methods is the fact that most were designed to model situations in developed countries and are merely adapted to situations in developing economies, resulting in limited interpretative success.

One of the most common transactions implemented in an e-ready environment is e-commerce/e-business. E-commerce transactions revolve around B2B, B2C, B2G, C2B, C2C and P2G, which form the largest proportion of the digital economy's business activities. The rapidly growing e-commerce industry has been given impetus by the open nature of the Internet in terms of limited regulation and the low costs involved, especially when compared to the more expensive form of EDI e-commerce that characterised the 1970s and 1980s. A number of factors are determining e-commerce adoption in the digital economy, including education and income levels, age, size of the enterprise (with large enterprises more likely to adopt e-commerce than small-sized enterprises), security of online business transactions, and the level of e-readiness. Various challenges constitute barriers to e-commerce growth and development, including issues of trust, policy and regulation, privacy, and skills and attitudes, to name a few.

SMEs are the backbone of most countries' economies and are expected - notwithstanding the barriers they face - to play a critical role in the digital economy in terms of their contribution to GDP and employment. SMEs are variously defined in different jurisdictions based on measures such as the number of employees, annual turnover, and level of fixed investment. Both developed and developing countries' economies revolve around SMEs, and they are generally seen as engines of economic growth worldwide. By and large, they constitute the majority of business establishments in most countries. Despite their importance to most economies,

SMEs face myriad problems, especially in developing countries. Some of these problems include: lack of access to credit, limited ICT skills, stiff competition from large enterprises, high cost of technology and connectivity, and limited international business networks. In order for SMEs to continue to occupy their enviable position in the economies of most countries, especially in the digital economy, the societies in which they operate and the enterprises themselves will have to become e-ready by obtaining modern technological infrastructure, competent ICT skills, access to credit, and other related factors.

The importance of the e-readiness of nations, and by extension SMEs, in the digital economy cannot be overemphasised. SMEs are increasingly adopting ICTs in their business transactions, especially the Internet and the World Wide Web, to reap the benefits of the digital age. There is a distinctive pattern of progression in the adoption of the Internet/ICTs among SMEs, from having simple static websites, for example, to developing more interactive sites where ordering and payment can be made online. Publicly owned SMEs are more likely to adopt ICTs than private but non-family owned SMEs, followed lastly by family-owned enterprises. While there are other factors that motivate SMEs to adopt ICTs in their business transactions, the most important is the perceived benefits of ICTs (e.g. ease of communication, advertising, creating portals, online banking, e-procurement and e-payments). Other factors that influence the adoption of ICTs by SMEs include the CEO's level of education, sophistication of online transactions, and costs of technology.

Internationally, SMEs in developed economies, such as those found in North America, Europe and parts of Asia, have attained a much higher level of e-readiness than those in developing countries. SMEs in developing economies face numerous problems, some of which have already been outlined. A further obstacle to SMEs' is their level of information readiness – their ability to determine their information needs and leverage the resources available, including ICT, to meet those needs. SMEs require various kinds of information to survive in the competitive global business environment, from start-up information to information about the day-to-day operations of the business. Most of the information sought may relate to legal assistance, market information (e.g. trends and prices), policy changes and legislation, technical assistance, and access to credit, among others. In order to satisfy these needs, SMEs require different information resources, such as the Internet, brochures, regional libraries, correspondence with colleagues in other enterprises, and participation in workshops and seminars. The information obtained is put to use in planning and policy formulation, understanding market trends, carrying out market intelligence surveys and more. In their endeavor to access relevant information for their operations, SMEs face further information seeking obstacles, such as inadequate skills, lack of access to information, and the inability to interpret the information received and put it to effective use.

Given that one of the main obstacles facing SMEs is their inability to determine their information needs, access the information, and evaluate and subsequently use it, the issue of capacity building in SMEs should be of critical concern to governments. Internationally, capacity building is increasingly being perceived by governments as vital to getting SMEs to participate in the global economy. Emphasis is being placed on education and training, particularly on including ICT as both a classroom tool and a subject. Within the European Union, the eEurope initiative encourages capacity building by investing in people skills, while in North America, the US and Canada have focused more on creating awareness among businesses of the benefits of ICTs. Canada has a Student Connection Programme that focuses on training university students as business advisors for SMEs. In the Caribbean, the government of Jamaica has invested heavily in education to meet the labour needs of this century. Within the South African Development Community's (SADC) member states, intention has also been mooted by regional governments to strengthen tertiary education at both regional and local level to foster the area's ICT and skills base.

Information literacy is critical to any meaningful education and training programme, especially with regard to the use of ICT. A number of models provide benchmarks that can be used to develop and offer an effective information literacy programme, especially in achieving the level of literacy required for the digital economy. Such models include the American Library Association (ALA) model, the British standard by SCONUL, the WSIS Declaration of Principles (which define the kinds of skills needed in the digital age), IFLA (provides standards for information literacy), and Mutula and van Brakel integrated e-readiness model with five dimensions of e-readiness that define information literacy competencies.

Globalization has been a critical factor in the emerging digital economy. Through the dynamics of globalisation, both large and small-sized enterprises are leveraging ICT to extend their market reach in international business environments. The Internet has increased business' trans-border data flow by facilitating access to market intelligence (knowledge spreads very quickly through the Internet) while also offering lower prices for transacting business. Multinational companies, and increasingly SMEs, can now deploy and mobilise resources and operations globally by using the Internet and related ICTs. With the increasing globalisation of business enterprises and the movement of human capital across borders, it is now possible for SMEs to successfully compete in global markets. These developments are making the rapid spread of e-commerce possible.

Countries in the developed world with high levels of e-readiness are finding it much easier for their companies to operate in the global business environment than enterprises in developing countries where the level of e-readiness is low. It is anticipated that with the globalisation of the digital economy, SMEs will find it prudent to enter the international market because they stand to benefit more from: reduced transaction costs, gaining competitive advantages, and using management supply chain networks that facilitate procurement, knowledge and skills sharing, and technology transfer. Enterprises aspiring to take advantage of the digital economy through ICT can learn from the developed world where most of the enterprises, by virtue of the high level of e-readiness of those countries, are already partaking in the global digital economy.

A number of countries worldwide and some international agencies are undertaking initiatives to enable companies operating in their jurisdictions and beyond to transact business online and effectively participate in the digital economy. For example, in South Korea, the government supports B2B and IT use in SMEs, while in Hong Kong the government does not tax profits from e-commerce. Elsewhere in Asia, the Mongolian government is driving its e-commerce programme from the Prime Minister's office. Within Europe, the European Union has also shaped its regulatory environment to cater for e-commerce. Furthermore, the OECD region has embraced the concept of learning organisations to adapt their organisations to the changing global digital environment by promoting a culture of continuous learning and process reengineering. In North America, the US Federal government is working with the UN Commission on International Trade Law to develop a model law that supports the common use of international contracts in e-commerce.

As concerted efforts are being undertaken globally to enhance global e-business environments, there are various challenges and issues that first have to be confronted and addressed in order to achieve a truly global digital economy. Issues of online security and privacy, for example, are exacerbated by the fact that the Internet does not have an effective central coordination and control system or mechanism. With whom, for instance, does one lodge a complaint if one's rights are violated online? Moreover, there is a general lack of awareness about the potential of technology and in particular, what e-commerce can offer, especially to SMEs. There is a need to educate business enterprises on the potential of ICT in business transactions. Governments should be encouraged to work through bilateral and multilateral arrangements to create cross-jurisdictional legal frameworks that enhance cross border e-commerce. In addition, the peculiarities of SMEs demand that the problems they face should be given special treatment if they are to effectively participate in the digital economy. The success of nations' participation in the Net economy is predicated on the level of their SMEs' e-readiness. Without the full participation of SMEs, this hyped economy may become the economy that never was.

THE WAY FORWARD

Both policy and managerial decisions are necessary on the part of government and SMEs respectively, to spur SMEs into active participation in the digital economy. SMEs are heterogeneous, undertaking a wide range of activities in different sectors. Whereas some SMEs are sophisticated users of the Internet and e-business, like larger firms, others are laggards. There is a need for a general policy differentiation to cater for such differences, since governments may not find it prudent to tailor policies to suit the requirements of individual SMEs. General policies that focus on delivery mechanisms while catering for the diversity of SMEs are desirable in order to maximise effectiveness. A policy's focus should also try to emphasise the general aspect of e-business rather than e-commerce, which is much narrower and more specialised, with limited applicability across the diverse range of SMEs. Such a shift in policy would cover R&D, planning, marketing and sales, invoicing, supply chain management, inventory, after-sales services, knowledge management, and finance, among others.

SMEs, on the other hand, need to conduct self assessments to determine whether they are able, among other things, to recognize what they wish to use e-business/e-commerce for; what benefits they stand to gain by delving into e-business/e-commerce or the digital economy; their preparedness in terms of skills, infrastructure, the legal environment and general business environment; aspects of e-business/e-commerce that they need to implement and in what critical areas of the business; the techniques needed to implement e-business/e-commerce; the associated costs and benefits; the resources and financial and human capital they would require; monitor their performance in the digital economy in order to take corrective actions; and recognize the various training needs. Payne (n.d.) summarises these well, saying that in order to determine how the e-readiness initiative should be delivered to the targeted SMEs, a needs assessment survey should be conducted. This assessment would help the SMEs define the target audience, determine their level of expertise and sophistication and their context, and identify what other programs are underway.

The take-up of ICT and access to the digital economy by SMEs can be enhanced by paying close attention to the availability of physical networks, such as Internet infrastructure and systems; economic networks (market and business organisations); resources (information and knowledge); and financial capital and policy (especially support and regulation). Alongside access to physical infrastructure, access to information and knowledge is the main challenge in the digital economy. The erosion of time-space-cost barriers and the decentralised network model makes ICTs extremely powerful in stimulating and/or accelerating growth, as they allow knowledge resources to break the chains imposed by old and rigid business organisation models, and favour creativity to explore new applications in different sectors. Countries and other entities require well-defined e-business strategies and plans if they are to obtain competitive advantages. The foundation for all e-business' readiness is based on modern technologies and access to those technologies through communications and information systems. Included in communications and information systems infrastructure are networking and computer hardware, underlying application software for e-business applications, and the applications representing automated business processes. SMEs that create and commercialize evolutionary new technologies directly are able to output new innovations ahead of their competitors (Jutla et al., 2002).

Partnerships are critical to the creation and spread of knowledge and the creation and diffusion of innovations because efforts of one party alone cannot ensure a healthy digital economic environment. Partnerships ensure that competitors do not replicate their innovations. Such partnerships are also important to SMEs in the digital economy because they bring together government departments, business organizations from the private sector, associations, and educational institutions. The importance of partnerships in business led OECD to suggest that governments should enhance the ability of national firms, especially SMEs, to partner with public research institutions and gain access to global research networks and markets (Germany, 2000).

Governments can play a mobilizing role in facilitating partnerships amongst private sector businesses, public sector departments and agencies, industrial research partners, public research players, universities and educational institutions, and industry associations to foster e-business-readiness. Governments can also nurture a conducive environment for the digital economy by supporting the setting up of research centres, helping e-business ventures in key national industrial sectors, facilitating the adoption of e-business by the SME sector, providing funding to e-business research, and creating additional programmes for higher education in e-business. Research institutes can be encouraged to set up e-business incubators to facilitate the transfer of innovation to the SME sector (Jultla et al., 2002). Encouraging the use of ICT by SMEs for various e-business and e-commerce applications depends on reliable and adequate network infrastructure. Encouraging the rollout and use of quality infrastructure at affordable prices can stimulate further uptake. Broadband connectivity also accelerates the contribution of ICTs to economic growth, facilitates innovation, and promotes efficiency, network effects and positive externalities.

The development of broadband markets, efficient and innovative supply arrangements, and the effective use of broadband services, require policies that: promote competition and liberalisation in infrastructure; network services and applications across different technological platforms; encourage investment in new technological infrastructure, content and applications; and promote technological neutrality among competing and developing technologies to encourage interoperability and innovation, and to expand choice. Extended public financial assistance to cover under-serviced groups and remote areas could complement private investment where appropriate (OECD, 2004). SMEs can also be encouraged to go online by providing them with regulatory infrastructure that ensures trust, security, privacy and consumer protection. This should be strengthened by cross-border co-operation amongst all stakeholders to achieve these goals. Of particular relevance for small-sized enterprises are low-cost online dispute resolution mechanisms between firms, and between firms and consumers.

A comprehensive summary of recommendations that would encourage SMEs to adopt ICT are proffered by OCED (2004), and they include:

- Moving beyond policies for basic connectivity and ICT readiness to facilitate more widespread uptake and the use of complex ICT applications and e-business by small firms
- Encouraging the rollout of affordable, quality broadband networks to underpin the competitiveness and growth of SMEs
- Developing and distributing digital content by expanding the commercial use of information about the public sector, education and healthcare
- Reducing ICT skills impediments to the growth of SMEs.

REFERENCES

Germany. (2000). *Benchmarking industry science relationships*. Berlin: Federal Ministry of Education and Research and the OECD.

Jutla, D., J. Dhaliwal, and P. Bodorik (2002)... Internet Research, 12(2), 139–164. doi:10.1108/10662240210422512

Mutula, S. M., and P. van Brakel (2006). An evaluation of e-readiness assessment tools with respect to information access: Towards an integrated information rich tool. International Journal of Information Management, 26(3), 212–223. doi:10.1016/j.ijinfomgt.2006.02.004

OECD. (2004). *Recommendation of the council on broadband development*. Paris: OECD.

Payne, J. E. *E-commerce readiness for SMEs in developing countries: A guide for development professionals*. Washington, D.C.: Academy for Educational Development.

About the Author

Stephen Mutula is a Honorary Senior Research Fellow at the University of Zululand. He is affiliated to the University of Botswana's Department of Library and Information Studies where, he is the Head of Department. He holds a PhD and Master's degree in Information Science, Postgraduate Diploma in Computer Science, and Bachelors Degree in Education from the Universities of Johannesburg (SA), Wales (UK), and Nairobi (Kenya), respectively. Prof. Mutula is the author of Web-based information management: A cross disciplinary text, and is also the co-editor of Information and Knowledge Management in the Digital Age: Concepts, Technologies and African Perspectives (Third World Information Services, Ibadan, Nigeria). He serves on several international editorial boards and has won several authorship awards, including the University of Botswana's Researcher of the Year Award for 2006 and 2007, and three Emerald Literati Club authorship excellence awards. He was also the recipient of a scholarly excellence award from the Standing Committee of East, Central and Southern African Librarians (SCECSAL) in 2000 in Windhoek, Namibia. At the time of writing this book, he was a visiting professor at the Department of Information Science, University of Zululand, South Africa.

Index

C

D

E

F

G

H

I

O

P

R

S

CPSIA information can be obtained at www.ICGtesting.com
Printed in the USA
BVOW061355101111

275532BV00006B/12/P

9 781605 664200